THOMAS LEVENSON

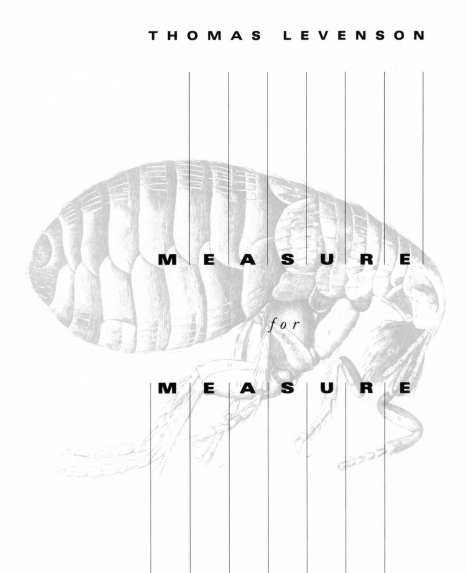

MEASURE

for

MEASURE

A Musical History of Science

T 16224

SIMON &
SCHUSTER
NEW YORK
LONDON
TORONTO
SYDNEY
TOKYO
SINGAPORE

SIMON & SCHUSTER
Rockefeller Center
1230 Avenue of the Americas
New York, New York 10020

Designed by Pei Loi Koay
Manufactured in the United States of America

10   9   8   7   6   5   4   3   2   1

Library of Congress Cataloging in Publication Data

Levenson, Thomas.
    Measure for measure : a musical history of science / Thomas
Levenson.
        p.   cm.
    Includes index.
    1. Science—History.   I. Title.
Q125.L44   1994
509—dc20   94–16282
                                                      CIP

ISBN: 0-671-78730-6

Portions of chapter 6 first appeared in *The American Scholar*, published by the
Phi Beta Kappa Society, as "How Not to Make a Stradivarius."

For Katha Seidman,

with love and a huge debt of gratitude.

In memory of Brian MacSwiney,

with whom I had hoped to debate much, much longer.

CONTENTS

Partly, this book is Bach's fault. Nineteen eighty-five was the three-hundredth anniversary of Johann Sebastian Bach's birth, a fact that led me, one winter afternoon, to the basement of a building hard by Boston's Old North Church (of Paul Revere fame). The building held a trade school, and its cellar was crammed with the pianos on which apprentice tuners learned their craft—old and tired cabinet instruments for the most part, though there was one beautiful mid-nineteenth-century cherry-wood piano made with an angled keyboard, the old cocked-hat design. I had come to ask what had seemed to me a simple question: What was this "well-tempered" keyboard that Bach used when he wrote his famous sets of preludes and fugues, now known as the *Well-tempered Clavier?*

Someone had cleared a space among all the larger instruments just large enough to set up a little hand-built harpsichord, and as I watched and listened, Bill Garlick, instrument builder, began to turn first one tuning peg, then another, and another. One note sounded, middle C, and then the second, the E just above middle C—the interval called a major third. The two together sounded clear and precise, what musicians and physicists alike call a pure consonance. Garlick reached for another peg, and I could hear the squeak of a string changing tension. He played the E and then G-sharp, the note one more third up the keyboard. Turn again, squeak again, and then the final third, up to the next C. Middle C to E, acoustically pure and perfect; E to G-sharp, acoustically pure and perfect; G-sharp (or A-flat) to C, pure and perfect. Then he played middle C and the C an octave above it;

and the two notes sounding together beat against each other and assaulted the ear like cats fighting.

There, for what I could make of it, was an answer: well tempering is a method of tuning keyboard instruments that can eliminate the howl heard in the octave built out of the three acoustically perfect major thirds my guide had played for me. After hammering my ears till I got the point, Garlick retuned those thirds, widening them slightly to make them somewhat dissonant, thus repairing the octave. Now the interval from C to C sounded smooth and clear, like a single tone. That stretching is the actual tempering of the keyboard, and depending on which notes within the octave are altered, can take many forms. Well-tempered instruments are democrats: every note on their keyboards has been moved slightly, distributing the distortion over the whole range, all to keep each octave absolutely pure up and down the full compass of the instrument. Bach's version of well-tempered tuning shares that attribute with the tuning system we still use; all the music we make with it is to some degree acoustically "imperfect," in a sense, out of tune—but we don't hear the subtle edge of discord, never having heard anything else.

That's the simple version. There is a longer answer, one that begins with the discovery of the musical scale and of the physics of acoustically pure intervals. The briefer one, though, satisfied me for the afternoon—except that I wondered what other choices Bach had as he tried to form his music within the constraints of the physics of sound, and which of them his predecessors had attempted.

In the spring of 1985 I traveled with some colleagues from a radio station in Boston down to the Marine Biological Laboratory at Woods Hole, Massachusetts. We were trying to produce a portrait of the laboratory, a remarkable place, founded originally to take advantage of the abundance of squid in local waters. (Squid have two giant nerve cells that run the length of their bodies—among the largest nerves found in the animal kingdom—which has made them uniquely valuable for neurobiological research.) I dropped in on Shinya Inoue, a cell biologist I had met once before, to ask what new territory he'd found lately. He took me to the little room—a closet, once—behind the room where he keeps his aquaria and he showed me the machine he had begun to develop when he first moved from Japan, right after the Second World War, rebuilding it continuously in the forty years since. More than six feet high (and still growing), it had become a Rube Goldberg collection of tubes and bits of glass, wires and cable connectors that Inoue had pieced together into what had become perhaps the most powerful light microscope in the world. With it he could look at objects below the theoretical limit of resolution for such instruments—bits of living cells smaller than half the wavelength of the light used to illuminate them. With each increment in the power and accuracy of his microscope, Inoue had seen what none before him could have discovered—the scaffolding of microtubules all cells build as they divide, for example, or the particular shape into which the DNA in the head of sperm cells coils itself.

But at that moment, Inoue was working on the machine, rather than on biological problems he could solve with it—adding on video cameras and computers that used once-secret image-enhancing programs (of the sort spy satellites employ) to extract detail out of moving images. The first video-microscopy movies had already been made—some of Inoue's colleagues had managed to catch sight of the chemical signals that travel in little packets up and down nerve fibers—but Inoue hinted that he would have better stuff in the bag soon.

I returned a few years later. This time, Inoue's microscope had grown so large it seemed ready to burst through its tiny chamber. He turned on one of the display monitors new since my last visit, and I saw what appeared to be a kind of balloon, with tubes or tunnels leading away from it. The largest, stretched diagonally across the screen, sported a fringe of small branches, each topped by a little berry-shaped ball. Using a joystick, Inoue began to twirl the image, looping me up, around, underneath it. Then he froze it with one of its branches center screen and pointed to the ball at the tip. It was, he said, another view of a nerve cell—and, he added, some people thought that those little balls were the place where memory resides.

Memory's library; looking at the image on the screen it seemed as if I could just reach in and unspool the hidden messages, unraveling what some mind had stored there. And looking longer, I marveled at how elaborate, how complex the process of discovery has become. With Bach and the mystery (to me) of the well-tempered scale, I encountered an idea that shaped how we interpret our experience, all that we hear in music, forever after. The physics of sound did not change with Bach—but how we use our knowledge of that physics has. With instruments like Inoue's microscope, I confronted the kinds of machine that force us to try to make sense of a world that changes continuously, with every increase in our ability to explore the universe we inhabit. Our instruments reveal what we must then try to fit into some coherent picture of nature. What do we know? How do we know it? What use do we hope to make of the knowledge we have gained? These are the questions that launched this book.

An instrument, a machine, contains within it a kind of archaeology of ideas: its design and construction reveal what its builders thought was important to try to do—what they wanted to get at or produce. It also may reveal (or create) that which molds or even overwhelms the expectations and assumptions behind its design. The word *instrument* is itself a rich one, and the way this book works turns on the use of its layered meanings. The Oxford English Dictionary gives a history of the word: It derives from the Latin *instrumentum*, meaning "provision, apparatus, furniture, an implement or tool, a document." It made its first appearance as a word in what was becoming English during the thirteenth century. Its first, and still broadest, sense is of an agent to perform an action—anything, as the dictionary states "that serves or contributes to the accomplishment of a purpose or end." The inevitable Shakespeare reference has it (from *King Lear*): "The Gods are just, and of our pleasant vices Make instruments to plague us." People can be instruments: "He sweares, As he had seen't or beene an Instrument to vice you to't"—Shakespeare again, from *A Winter's Tale*.

The first narrower definition of the word the OED acknowledges is the familiar one of a material tool, an implement used for the "accomplishment of some mechanical or other physical effect." Shakespeare once more, this time from *Romeo and Juliet:* "Here is a Frier and Slaughter'd Romeo's man, With Instruments upon them fit to open These dead mens Tombes." Commonly, though, the word came to refer to implements designed for tasks more delicate than grave robbing. From almost the first examples of English vernacular writing, the word refers especially to scientific instruments. Chaucer used it that way in 1391, recording the "Conclusions apertyning to the same instrument" in an account of work with an astrolabe.

From pleasant vices to the tools that measure heavens—the word travels well, and farther still: English is itself an astoundingly flexible, malleable and subtle instrument (in the OED's first sense)—and even as relatively straightforward a concept as that of an instrument has evolved shades and shades of meaning over the centuries: it has described body parts—"the instrument of breathing"—deeds, charters, weapons, and more. But along with the tools scientists use, the most important other apparatus that can be defined as an instrument is that "contrivance for making musical sounds." Instruments are the machines that make music—as in one last, not terribly poetic Shakespeare quotation, from *Othello:* "Are these, I pray you, Winde instruments?"

This book turns on the element that links the instruments that serve either scientists or musicians. For both, instruments serve as extensions of human capacities. Those extensions augment our own abilities, but of necessity do so narrowly: a hammer increases the force we can bring to bear on a nail but is far less dextrous than our fingers. A microscope extends our sense of sight, bringing into view the very small, but it confines our vision to an ever more confined field: we see more and more of less and less. A harpsichord allows us to pluck particular pitches, but we are confined to the notes its strings define, and to the tone, the timbre, that the act of plucking a taut thread creates. Each instrument, for both music and science, is built to a particular purpose: to perform one measurement or another, beyond the capabilities of unaided human senses, or to make one set of sounds or another, as we choose. Each choice is both an affirmative statement—this action is important, this action serves our purpose—and a demarcation—this action is more important, is more suited to our purpose, than the ones we cannot now undertake, having chosen to pound, to look, to play.

Music and science have been intertwined in Western thinking from the moment of their shared origins, of course: the first even vaguely scientific theory of the universe was a musical one, Pythagoras's arrangement of the planets on the scaffolding of his musical intervals, with every heavenly body sounding out its note in what became known as the music of the spheres. But the fact that both musicians and scientists do their daily work with machines they use in similar ways suggested to me that the link between their endeavors runs far deeper than that of metaphor. At some level, it seemed to me, the two disciplines think in similar ways—and examining one could shed light on the other. I decided to look more closely at a few extraordinary instruments as they emerged in each field as a way of piecing together a new perspective on the evolution of scientific enterprise: I could ask, not what people had thought, but how they did their thinking.

In doing so, I am taking one slice out of a larger story. Perhaps the most important ill-told tale in the history of Western science is of the extraordinary influence the rest of the world had on the development of what used to be thought of as a purely European invention. Both the ideas and the technology vital to the advancement of science and the intellectual triumph of the scientific revolution flowed in from outside Europe. The probable Chinese origin of eyeglasses, mentioned briefly below, is but one token of the transfer of inventions from East to West. Similarly, the story of alchemy is an instance of the vital significance of Europe's contact with an Islamic civilization that produced enormous intellectual riches of its own and transmitted both Europe's forgotten Greek heritage and the lode of scientific advances developed farther East again, in India and beyond. There is, in other words, much more to tell than has yet been told about how a relatively small and provincial set of people began to examine their world in a way that would lead, ultimately, to a modern scientific worldview. In the book that follows, though, my focus remains primarily on the Western story, what happened within a single culture, the culture that eventually became the ubiquitous scientific one of our own century. This is not to deny the non-European connection, only to ask a slightly different question of history: What do we, that history's heirs, actually seek to find out with our invention of scientific ways of knowing?

And so the book that began as a stray question about Bach, and a glimpse of what might have been a memory, has become an account of how scientific thinking developed from its Pythagorean origins to the present day. The story breaks down into three parts—based not on the usual sequence of discovery but on the kind of question being asked in each major era of science. Part one explores how science works when it is an inquiry into the reason why: given the order to be seen in nature, what caused that order to form—why is the world as it is; what does nature tell us of the mind of God? Such questions dominated science from the Greeks through the Middle Ages—but questions only shape, they do not determine the answers. Part two looks at what happens when results did not match the expectation that every scientific advance would yield further evidence of perfection, of God's order. Once it became clear that what happened in nature did not fit a preordained sense of what ought to be out there, the function of scientific inquiry shifted, and the dominant question became What?—what occurs; what relationships exist; what phenomena does nature actually display;

what are the details, and how are they arranged? The scientific revolution of the seventeenth century made the asking of such questions systematic and established the formal, intellectual apparatus to produce the right kind of answers—but it retained an assumption from the older era: nature can be known, the book of nature can be read completely, from cover to cover. That assumption held good for a long time, but by the end of the nineteenth century, the astounding success of Western science in discovering ever more detail in nature began to break it down. Part three picks up the story at the point of fissure between Newtonian science and the modern version, when the fundamental question being asked of nature shifts again. Instead of trying to discover what exists, we now attempt to investigate only one part of nature at a time. Instead of studying the whole in its pieces, we investigate piece by piece, asking of each phenomenon we observe in turn, How does it fit? How does its behavior resemble that of some other aspect of nature we have already encountered?

Why, what, which (and what are the probabilities that what we see is true)? What we have sought to find out, what we think we can learn about our surroundings has shifted continuously over the history of science; what follows is an account of how that occurred, and of where that process has left us now: what kind of understanding of nature we now believe we possess. The stories of the instruments, the scientists and musicians who used them, each form a kind of microcosm, a miniature of the book as a whole. Science itself is a macrocosm, a kind of meta-instrument, a tool we have built over millennia to pry loose some sense of what our world contains. And like the actual instruments scientists have built to aid them, science itself has, throughout its history, both narrowed and sharpened its focus. We lose and gain; we know more and more—and learn yet more of what we cannot know, cannot learn, using this peculiar kind of reasoning, this scientific investigation of the world. The story comes to rest here, with awe at the power of the human passion to ask, and a greater reverence for the artistry that constructs an answer—some answers—amid all that we seek to understand.

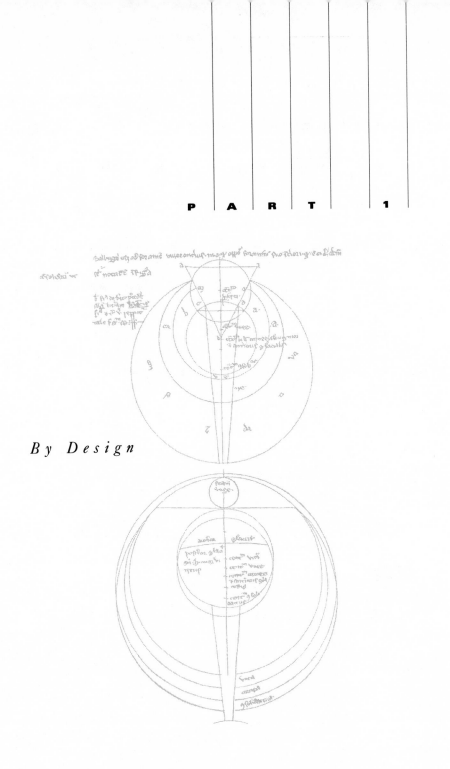

By Design

SCIENCE AND ARTS ACADEMY

Chapter 1

A
PERFECT
ORDER

The Madonna holds the infant Jesus on her lap.
The baby reaches out to his right, placing a ring on the finger of a demure beauty, before whose feet lies a wooden wheel—Catherine of Alexandria, saint. The two Johns, Baptist and Evangelist, watch on while an angel plays the wedding music. In the background one can see through the windows to the streets of a town, a large one, possibly those of the town in which this altarpiece was painted: Bruges in the Burgundian provinces of the Netherlands.

Hans Memling left his native Germany for the Low Countries around 1467, already a master painter. Bruges was then at the height of its wealth and glory, a center of the arts—its population then three times the size it would be by 1900. Jan Van Eyck had painted there his sharply focused images of life and faith. Memling's was a different art, more peaceful and still. In this, his *Mystic Marriage of Saint Catherine,* there is only a hint of the rest of the legend: beatings, imprisonment, torture (on the wheel— hence her emblem and the name of the children's firecracker), and finally beheading. Memling's vision is one of comfort—a good room, well appointed, and the joyful consummation of belief. His Bruges was a fine city, smiled upon by heaven. His saints are richly dressed, placid, beautiful. Monied donors commissioned his work to adorn the houses of the church—the Saint Catherine altarpiece remains in Saint John's Hospital. "Gloria in Excelsis Deo" ("Glory to God in the Highest") that they did live in such times.

There is a delightful detail in Memling's picture. In the background, the angel plays. Her instrument is a portative, a tiny pipe organ, held in the crook of her arm. Her left hand is out of sight, pumping the bellows, while her right fingers a keyboard that runs for just over two octaves. She (for the sexless angel is here represented as a young girl) is a creature of heaven, playing the music of divine love, on an instrument built to an ideal of perfection. There is harmony in the scene; harmony between heaven and earth; harmony in the sounds that the artist allows us to hear within our minds' ear, issuing from the double rank of pipes, sounding to an angel's touch.

Little organs recur in medieval and early Renaissance art, often in the hands of angels. There is some evidence that portable organs existed in Europe as early as the tenth century. Later, they served as a convenient motif for painters, conveying an image of luxury—even small organs were not cheap—and sanc-

Memling's "The Mystical Marriage of St. Catherine," complete with an organ-playing angel.

tity: small organs evoke large ones, the heralds of the church. And they were beautiful, pipes and woodwork delicately figured, gracefully proportioned. For Memling the instrument was an easy, common, clearly understood symbol, and he used it more than once in his work. In time, though, portable organs disappeared. Their passing marked an end and a beginning: the loss (that we still perceive) of an idea of perfection, the start of a renewed attempt (still under way) to replace it, to come up with a picture of the world as complete as that reflected in the panels of Memling's masterpiece.

Begin here: an angel sounds a tone, and then another, an octave apart. The modern definition of an octave uses the notion of frequency: sound travels through a medium—air, for example—as a wave; two notes form an octave when the wave sounding the higher pitch oscillates at twice the frequency of the lower tone. The octave, thus defined, is an abstraction, an arithmetical construct imposed on the physics of waves—an invention. But there is an intuitive sense of the octave that is as old as humanity: the register of adult male voices is, on average, about an octave below that of adult females. A man and a woman singing together would have discovered the pleasant sound of a doubled melody ringing out at an octave's remove.

Within the confines of the musical scale, the ancients constructed a theory of everything. The modern, Western scale descends, in part, from the holy arithmetic of Pythagoras, his followers and his antagonists. The discovery that transfixed the Pythagoreans was that the octave and other intervals that like the octave sounded harmonious and smooth occurred not simply by chance but as if by design. Pythagoras was credited in antiquity with the realization that there was a deep connection between mathematics, numbers, and sound: he discovered that the fundamental intervals in music were created by the perfect ratios of the lengths of string or pipe used to generate the notes.

In legend Pythagoras stumbled across the numerical ratios that define musical intervals while walking past a blacksmith's shop. As his devoted follower Nichomacus tells it: "Once upon a time . . . by miraculous chance he [Pythagoras] walked by a smithy and heard the hammers beating out iron on the anvil and giving off the sounds that are the most harmonious in combinations with one another. Delighted, therefore, since it was as if his purpose was being achieved by a god, he ran in to the smithy and found by various experiments that the difference of sound was consistent with the weight of the hammers, but not with the force of the blows, nor with the shapes of the hammers, nor the alteration of the iron being forged." As the medieval music theorist Boethius takes up the tale, Pythagoras discovered that "those two [hammers] which gave the consonance of an octave were found to weigh in the ratio 2 to 1. He took that one which was double the other and found that its weight was four-thirds the weight of a hammer with which it gave the consonance of a fourth. Again he found that this same hammer was three-halves the weight of a hammer with which it gave the consonance of a fifth." Pythagoras then ran home, according to later Pythagorean hagiographers, and repeated the experiment on his own, fixing a stake into the wall of his house, to which he tied four identical pieces of string. He then hung from each string a weight comparable to one of the hammers he heard at the blacksmith's shop, and plucked. Harmony reappeared, in ratios identical to those he'd observed in the beating of hammers against metal.

This particular account is certainly legend rather than fact. Pythagoras himself may have had a keen experimental bent, but clearly his followers did not. Hammers of different weights striking the same anvil give off the same tone, at different volumes—it is the bell, not the clapper, that sounds the note. And the strings that Pythagoras allegedly played to discover the harmonious conjunction of number and pitch would, in fact, sound horribly dissonant: a change in tension does alter the pitch of a string, but the pitch varies not directly but with the square root of a change in tension. Simply doubling the weight at the end of the string supposed to give off the octave would produce a note with a pitch in the ratio of 1.414 to the 1 of the original tone— which could sound harsh and clashing. To get the doubling of frequency required for an octave, Pythagoras would have had to multiply the weight hanging from his string by four, not two.

So the tale is a myth. Pythagoras is a strange and shadowy figure anyway, represented as almost godlike in the acuteness of his perception. Similar stories exist in other traditions, telling of the origins of music itself, and the Pythagoreans may simply have adopted an older tale to glorify their sage. But though the tale is false, the moral is true.

In reality, Pythagoreans later used an instrument called a monochord, a device with one string strung against a body, divided into two lengths by a bridge, like that on a violin to investigate musical ratios. The bridge of a monochord was movable,

Medieval drawing of the legend of Pythagoras, showing the master listening to the sound of hammers striking an anvil.

permitting the Pythagoreans to divide the string into the right lengths to discover each of the essential ratios ascribed to Pythagoras's impossibly fortunate stroll past the blacksmith. The ratios thus discovered do work: two strings, one precisely 3/2 times as long as the other will sound together in a perfect fifth, as will strings matched in the correct ratios for the octave or the fourth. Almost certainly, the fundamental arithmetic of the musical scale was built up of the relations between the numbers 1, 2, 3, and 4, a system identified by a living man called Pythagoras during the sixth century B.C. This discovery of the direct relationship between the pitch of a note and the length of string or pipe that produces it remains the oldest mathematically expressed law of nature. The Pythagoreans were not scientists; they sought magic in numbers. But still, here is where science begins.

Almost three thousand years separates us from the extraordinary sense of revelation Pythagoras must have felt at the moment he recognized the meaning of what he was hearing. We are accustomed now to the notion that experience—the roar of a church organ, a breath of air, the motion of the moon across the sky—is governed by the rules we call laws of nature. These are universals: the same mathematical abstraction can describe the behavior of an apple falling from a tree and the orbit of a star around a galaxy. But for Pythagoras no such certainty could have existed, until he plucked his strings, heard his tones, and recognized the relations of number that governed absolutely the resulting harmony. For the first time the ephemeral evidence of the senses could be accounted for by an idea that would hold true for anyone, at any time.

Such perfection was intoxicating. From the sounds of a monochord, the Pythagoreans deduced a universe. The planets moving through the heavens gave off sounds, "the music of the spheres," which exemplified the perfect organization of nature on the largest scale. The original thought, as Aristotle records it, was that the great bodies of heaven could not move without making noise, but later traditions created a musical cosmology. The regular motions of the planets suggested ratios of number, of intervals; the suggestion was enough to produce the image of celestial bodies ringing harmony throughout the universe. Such glorious music would yet be imperceptible—continuous, unbroken, it surrounded the listener from birth to death, with never a moment of true silence to act as its foil. The Pythagorean tradition has it that the master himself, alone among mortal men, could hear the perfect harmony of the spheres. But others, their senses less acute, could still discern the organization of nature in the patterns set out in the musical scale. Heaven and earth could be seen as a "cosmic monochord"—as the Englishman Robert Fludd, writing at the far end of the Pythagorean tradition (in the seventeenth century), would represent it. God's hand stretches the string of monochord, which passes through two octaves, from a high G in the sphere of the angels through the solar system to the sun at the middle G, down past Venus, Mercury, and the moon, through the elements, fire, air, water—on down to the resonant bottom G of earth itself.

And what of Memling's angel, up there in the realm of a high G? She is a symbol of perfection, of course, a creature of heaven. But she plays an instrument of this world, the organ, making music that saints, at least, may hear. The tones of her organ are fixed in the harmonies built into the lengths of the pipes she sounds. We may take her, for the moment, as an emblem of the glorious unity of God's creation, a Christian

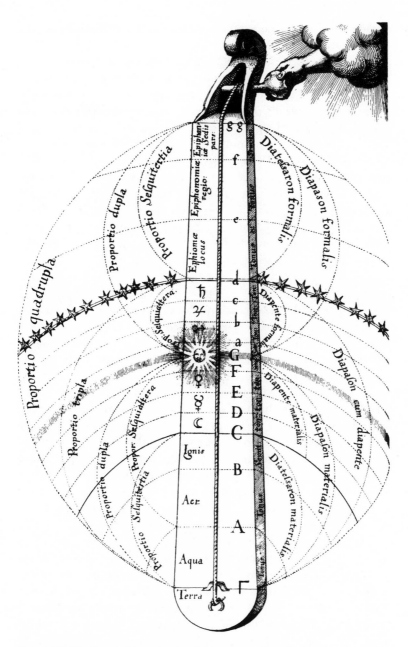

Robert Fludd's Cosmic Monochord—all the universe from God turning the tuning peg through the heavens down to earth.

version of Pythagoras's credo that all the world rings to the same music.

As in a sense it does. Leonard Bernstein in his Norton Lectures at Harvard in 1973 spent much of an evening trying to recapture something of the first discovery of the musical scale. Using a piano rather than a monochord, he performed the following demonstration, easily replicable on any reasonably resonant instrument. Strike a note—C′, as the note two octaves below middle C is written. Hold it, listen to it, get a feel for it. Now hold down the note C, one octave up the keyboard. Just depress the key, don't sound the note. This lifts the hammer off the strings, undamping them and allowing them to resonate freely. Now strike C′ hard; release the key, cutting off the note, and listen. Faintly that next C sounds, the higher string ringing at the strike of the lower in what is called the first overtone of C. Now undamp the G above the C, pressing down on that key without striking a sound; play the original C again; and listen for the sound of the fifth, the second overtone. Go up a fourth, to middle c, play the tonic again, and listen for the third partial. It is almost impossible for humans to hear further partials, but the higher overtones are there. The next falls on e, the third, above middle c. Above that the partials crowd in closer and closer together, with what is called the harmonic series cumulatively providing that sense of color, the richness of sound that surrounds the pure tone of the original C′. (There is another factor that gives depth to a piano's sound: Its notes are struck on three strings, each just slightly out of tune with the other, which together produce a complex, powerful tone. But the overtones still ring out, creating the full envelope of every note played on the piano.)

In the modern account of the wave dynamics of sound, the overtones occur because the original string vibrates in a very particular pattern. There is the wave formed by the whole string—the loudest sound heard. But the string will also vibrate in parts, so long as those parts add up to the length of the whole string. Thus the first string vibrates as if it were cut precisely in half—the old ratio of 2:1, the octave, or first overtone. Next, it divides into three, producing the ratio of 3:2 over the first overtone, the G, or the fifth; likewise in four parts we get the ratio of 4:3, and hear the second C above the fundamental. And so on— 5:4; 6:5; 7:6; 8:7; 9:8—an infinite (if soon inaudible) series of tones. All those vibrations set up sympathetic vibrations on the strings of the piano that correspond to particular overtones, and hence produce the faint echo one can hear at the piano bench.

For Bernstein the harmonic series led further. The first six in the series of overtones, the octave, the fifth, the fourth, the third, and the next one, the second, create a five-note "pentatonic" scale—the scale that can be heard by just playing the black keys on the piano. (For the black keys: C-sharp to C-sharp is the octave, the first partial. G-sharp is the fifth, F-sharp the fourth. A-sharp is a minor third below the higher C-sharp, and D-sharp is the second—six partials, five intervals, five black keys: the pentatonic scale). That scale, Bernstein suggested, is amazingly widely used, heard in gospel hymns and symphonies, ragas and highland ballads. And the series exposes why the Western scale has twelve tones: take the fifth, the second overtone, and keep adding fifths on top. Go from C to G; then from G to D; from D to A, and so on. Twelve different notes come up until finally the round of fifths lands at the thirteenth turn on C, where the cycle may begin again. The perfect fifths divide the perfect octave and create the scale of twelve notes that still forms the scaffolding for all Western music.

And as for Saint Catherine, Memling has given her angel a portative in which these developments can be clearly seen. There is a twelve-note scale breaking up the octave, and the five "black" keys—here just pegs elevated above and behind the first rank of keys—can be seen as well. In fact, both of the crucial discoveries of Pythagorean music theory—the whole-number ratios and the cycle of fifths creating the scale of twelve notes—were in place at the time the organ appeared: it was the organ that brought together the technological developments of applied Greek science and the abstractions of Greek theory. Its name derives from the Greek *organon* (tool or instrument) and it was as a tool that it was first conceived.

There was such a beast—the first organ. The instrument was not developed slowly over time; it was built, invented, by one man, Ktesibios of Alexandria, who flourished around 270 B.C. Ktesibios was an engineer, specializing in the study of pneumatics. One of his inventions was a device to produce "intermittent bird song." It worked by regulating the flow of water into a closed cistern. As the cistern filled, air was forced out a pipe in the top, which led to a whistle hidden inside a figurine of a bird.

Ktesibios built several versions of this toy, but it had limitations that must have been frustrating: the air pressure produced in such a device is very low, so the bird song must have been faint indeed—and every time the cistern filled up completely, the music stopped. The invention was hardly original, anyway— Archimedes (who discovered the concept of specific gravity while sitting in his bath) is said to have built his own version, which Ktesibios probably knew. Ktesibios's next effort, though, was a breakthrough, the creation of an instrument whose essential features have remained unchanged to the present day.

What his bird whistle lacked was sufficient air pressure and a steady source of pressurized air. It also couldn't be controlled— whatever whistle or whistles were built into the device played when there was air and shut up when there wasn't. They contained one element that the organ would require—the whistle or pipe. Ktesibios added the three others that now define the instrument: the pump, to supply air; the wind chest, to distribute it properly; and the manual, to allow a player to control which pipe or pipes would sound and which would not.

Of them all, the pump represented Ktesibios's greatest technological achievement. He had already invented a double-barreled cylinder pump, used apparently to help fight fires. It was made of bronze, highly polished, with a bronze piston fitted tightly within the body of each cylinder and a handle attached to the piston rod. The pump could suck up water from a well and force it into a holding tank, which would compress the air within that tank. Release a valve, and the compressed air would expel the water from the tank through a tube or hose onto a fire.

For his organ Ktesibios modified this arrangement to create a continuous source of air compressed to a roughly constant pressure. As the Roman author Hero reports from the first century B.C., Ktesibios used three major components to build his wind source: a single-cylinder air pump, a large cistern mostly filled with water, and a smaller vessel, called the pigneus, fixed to the bottom of the cistern to serve as a regulator. The air pump had an intake valve, to draw air into the cylinder, and an outflow valve, which led to a conduit carrying the compressed air down to the pigneus, a hemisphere of brass with an outlet at the bottom through which water from the cistern could flow. The compressed air forced water out of the pigneus, while flowing up through another conduit toward the organ and its pipes. When the pump relaxed, the outtake valve shut, air pressure in the pigneus dropped, and the water forced out by the pump could reenter the smaller hemisphere. That flow of water into the regulator chamber compressed the air there again, and with the valve to the pump shut, there was no place for the pressurized air to go but back up into the organ. That provided the continuity—no matter whether the pump was on its upstroke or its downstroke, air from the pigneus remained under fairly steady pressure.

The other three parts of the Ktesibios organ were somewhat simpler. The air from the pump entered a wind chest, or reservoir. The pipes rested on the wind chest, mounted in holes. Beneath each pipe was a slot, or drawer, with a hole in the lower slat opening to the main cavity of the wind chest below. The slots were blocked by pieces of wood, or sliders, which were perforated by holes that precisely matched those at the bottom of the pipes. The sliders were controlled by the keys of the manual, or keyboard; when a player pressed a key, the hole in the slider aligned with the opening of the corresponding pipe. Air from the wind chest would then enter the pipe, and a note would play. The keys were attached by cord to springs, and when the player released a key, the spring would recoil, pushing the key back up, which dragged the slider to its original position and cut off the note.

That was it: the first organ, called a *hydraulis* in recognition of the crucial role of water in producing its sound. Except for the use of hydraulic technology to serve as a regulator, however, the basic design of Ktesibios's organ is recognizably similar to all subsequent organs—every one possesses a wind source, a wind chest, a keyboard and the pipes. Everything else hung onto an organ is in some sense bells and whistles. The core of the instrument has remained unchanged for more than two millennia.

The use to which the organ has been put, however, has evolved continuously since that moment. There is a legend that Ktesibios's wife, Thais, was the first organist, which, as one writer put it, "has an apocryphal air about it." Ktesibios himself seems to have had purely an engineer's motivation: the complexity of the machine, and the opportunities the problem afforded, attracted him more than any love of music. From his perspective, the hydraulis was a tool with which to explore the practical implications of the new science of pneumatics, the study of the behavior of compressed air. He is credited with war machines—he is reported to have built a compressed air-driven catapult—various pumps, and, in a slightly different technological tradition, water clocks. The hydraulis demonstrated his grasp of the most sophisticated scientific and technological knowledge of his day—he clearly recognized in the design of his regulator, for example, that air is compressible but water is not—and it displayed his mastery of precision engineering, with the fine machining of the pumps and the delicate work needed to build the wind chest, sliders, and keyboard.

It was the technological sophistication of the machine that attracted the notice of ancient commentators, certainly. Jean Perrot, whose book *The Organ* is one of the best sources of information on the early history of the instrument, reports that Ktesibios, a master of "the science dealing with the construction of marvelous appliances," was placed in the same class as Archimedes by the Roman writer Pliny the Elder and the Alexandrian mathematician Proclus. Ktesibios himself, along with several of his contemporaries and successors, some of whom appear to have been his students, participated in one of the last great flowerings of Greek science, out of which grew a major body of work on the theory of gases. They knew that air was a material substance; they argued that it was made of discrete chunks of matter "composed of tiny particles, light and generally invisible," as Hero of Alexandria wrote. Philo, one of Ktesibios's near contemporaries, even performed an experiment on the burning of gases similar in design (though not in intent) to the one that Lavoisier was to use two thousand years later to demonstrate the consumption of oxygen during combustion.

In this community, among men who performed some of the first significant experimental inquiry, the hydraulis must have appeared as a kind of virtuoso performance, a public demonstration of the remarkable feats inspired by the new knowledge being created in Alexandria. The actual sounds the hydraulis produced were less important than that it could make sound at all. Pythagorean notions of divine intervals mattered less than the precise arithmetic of the engineer's compass and ruler; numbers counted for more than Number. The contraption worked: the sounds of flutes and trumpets spoke forth, audible above the creak of the pump, and the burble of water. It was, literally, a marvel—Athenaeus, a Greek living under Roman rule in the second century B.C., describes a feast at which all manner of novelty was discussed: "The sound of the hydraulis was heard close by. So pleasant and charming was it that we all turned towards the sound, fascinated by the harmony. Then Ulpian looked at the musician Alcides and said: 'You who are the most musical of men, do you hear that wondrous symphony that caused us to look around in rapture?' "

A medieval drawing of musical instruments in use, showing a hydraulis in action.

That was how the hydraulis was heard, as an amusement in the midst of dinner, a diversion, a kind of carnival instrument. The first reference to it actually being played honors the skill of one Antipatros, performing at the Delphic games of 90 B.C. The Romans readily adopted the Greek invention. Nero was known to be an organist, and organs provided musical effects for the theater, and entertainment at banquets and circuses, even gladiatorial contests. The oldest archaeological remains of an organ were found at Roman Aquincum, near Budapest, dedicated in A.D. 228 to the college of weavers there—evidence, according to Perrot, of the extraordinary popularity of the organ within the Roman empire.

But if Ktesibios's masterpiece soon became commonplace, a plaything for dilettantes like Nero or a perquisite for the worthy guildsmen of Aquincum, it possessed from its inventor's day its essential attribute as the most sophisticated product of what we would now see as science. It retained from its origins a double character: it captured in one machine the elements of what we would now divide into the realms of experiment and theory. In itself it provides one of the first examples of technology inspired by the study of physical questions. Ktesibios's program is recognizable, was recognized by his peers, as a sustained application of reason to the understanding of nature. The organ carries within it the entire story: problem and solution, and even some sense of the method with which that problem, such problems, may be solved. Ktesibios's writings are lost, but his instrument, unchanged in its essentials, remained a text in wood and metal for whoever encountered it over time.

At the same time, it provided an interpretation, fixed again in ranks of metal pipes, of one of the fundamental theories the ancients possessed. The construction of the Western musical scale presented certain problems to the Greeks (and others, later, as will be discussed), but the underlying verities of the perfect consonances were built into the lengths of pipe. The organ gave its listeners the direct experience of concord and demonstrated for all to see that this sensation of perfection could be directly related to physical reality: the pipes were real objects, their lengths set and visible. A listener could hear the perfect consonances and see their cause; the connection between abstract mathematics and the real world was made. Little else remains of the details of Greek scientific thought, but this does. Let those who doubt come and listen.

Some did, and those who listened to the sounds of the hydraulis must have heard what the Pythagorean Philolaus proclaimed: "The nature of number and harmony admits no falsehood. . . . But in fact number, fitting all things in to the soul through sense-perception makes them recognizable and comparable with one another." There is a tantalizing air to such passages, for the Greeks did so much, and so much was later let go. Ktesibios and his near contemporaries did in fact lay the foundation upon which science as we understand it would be built. But their followers forgot much of what had been achieved, and their accomplishment virtually vanished within a few centuries.

With hindsight, one can see the disaster beginning in the last years of the Roman republic, when rogue generals embarked on free-lance missions of conquest. Lucullus, himself educated in Greece and a lover of the older culture, led the invasion of Asia, the kingdoms of Bythnia and Pontus. He set out to capture the wealthy Greek cities there, the outposts of Hellenistic civilization on the Black Sea. When Amisus finally surrendered in 71 B.C., the legions gorged themselves. Lucullus himself ran among his troops, trying to restrain them, but failed, and the city fell to the sack. The tone was set. In 48 B.C., Julius Caesar invaded Cleopatra's Egypt, and when suddenly trapped within the palace at Alexandria sought to destroy the fleet threatening him. He set fire to the ships, but the blaze blew out of control, onto shore. It swept through the city, and though the great library was spared, 40,000 books in a depository near the docks were destroyed. The Dark Ages are said to have begun in 410, when Alaric the Goth conquered Rome itself, but long before the babarians came, the smoke from the fires of the cities put to the torch by Roman armies carried the first hint of how much could be forgotten, how much rejected. With the rise of Christianity, and the doctrines of faith, Western Europe had an alternative to Greek reason: truth could come from God, not Number. Christians did not love mathematics; Euclid, geometry, algebra were simply abandoned. In the twelfth century Western men would have to learn from the Arabs what Greeks had discovered long before, as well as a great deal that had been discovered in China, India, Persia and beyond.

And music, as the Greeks used it to interpret nature, was unpopular with the divines of the early church. Systematic scientific investigation mattered considerably less than contemplation of divine revelation, and the sense of beauty, the emotional impact, that music could inspire was downright distracting. Organs disappeared from Western Europe with the fall of Rome, and it is not simply barbarism that can account for the loss, for organs were still known at the court of Byzantium and among the Moors. Rather, just as the organ embodied specific accomplishments of the Greek mind, its banishment from the Western tradition and its subsequent return reveals where priorities lay for many years.

Saint Augustine, bishop of Hippo in what is now Algeria, a thinker and a man of some humor, was more honest than many. Writing in the early fifth century, he found music in any form suspect, but, he allowed: "Now when I hear sung in a sweet and well-trained voice those melodies into which your words breathe life, I do, I confess, feel a pleasurable relaxation. But," he added, "this bodily pleasure to which the mind should not succumb through enervation, often deceives me . . . in these matters I sin without realizing it. . . . I pray that you, my Lord God, will hear me and look down upon me and observe and heal me. In your eyes I have become a question to myself, and that is my infirmity."

Augustine and the early church fathers clearly recognized the danger: the power that music has to elevate, exalt, to speak directly to the emotions might speak to the greater glory of God, but it could just as well be the devil's snare, to trap the unwary into pursuing pleasure for its own sake. Augustine himself explicitly condemned only the music of human voices, but his proscription was clear enough, and no organs appeared in church until the ninth century at the earliest—and then only after they had been reintroduced to the West for a completely different purpose. The first organ to reach Western Europe since the ultimate sack of Rome in 476 came as a gift from the Byzantine emperor Constantine V to the Frankish king Pepin in the year 757. The gift was so marvelous that, as Perrot reports, it was the main, perhaps even the sole, entry for the year in the various monastic chronicles, as "an instrument never before seen in France."

This organ almost certainly made use of the major technological improvement in organ building since Ktesibios's day—the use of pairs of bellows, rather than a water-regulated pump, to supply a constant flow of pressurized air. But the organ remained a foreign device, a mechanical wonder that the lord of the Eastern Roman Empire could use to remind the barbarian kings of the west of Greek preeminence. Such arrogance could not sit well with Pepin's son, Charlemagne, crowned emperor by the pope of Rome, nor with Charlemagne's son, Louis, styled "the pious." In 826, a priest, Georgius, a Venetian, came to Louis's court with a proposition: he would manufacture an organ, then and there. Venice was nominally under the protection of Constantinople and was, even then, the commercial center that linked Asia to Italy and the European hinterland; whether Georgius learned his craft at home, or had traveled in the east and studied the art of organ building closer to its roots is unknown. But on his arrival at Louis's court, the Carolingian emperor put him straight to work, building at that far end of the world an instrument "in the Byzantine style."

Even from the fragmented records that remain there is an air of haste in the drive to build that first post-Roman organ at a barbarian court, an urgency that seems to have little to do with an inordinate love of music. The motivation was partly technological, perhaps, but truly political: to display power through the demonstration of technological mastery. To possess an organ was to possess civilization. The Byzantine emperor had it; the Carolingian meant to get it. The court poets were quick to recognize the organ's symbolic meaning. Ermold le Noir's epic, written in praise of Louis, proclaimed:

> *Even the organ, never yet seen in France,*
> *Which was the overweening pride of Greece*
> *And which in Constantinople, was the sole reason*
> *For them to feel superior to thee—even that is now*
> *In the palace of Aix.*
> *This may well be a warning to them that they*
> *Must submit to the Frankish yoke.*

That's a grand claim for a ruler whose reach did not extend much beyond the Rhine, to say nothing of the Bosphorus, but the implication is clear. As the eastern and western halves of what was once one world split asunder, the two sides began to compete for the legacy of Rome, the right to assert supreme leadership. The organ could be seen to fill the same role as, say, the race to the moon: landing Neil Armstrong at the Sea of Tranquillity did little to affect the balance of power, but it was a convincing display of technological prowess, and was clearly recognized as an emblem of superpower status. So it was with the organ: if you could build one on your own, you were a player, at least. The organ required sophistication to build, and it allowed its possessors to boast of their growing power, symbolized by an instrument that could produce the marvelous roar of sound no unaided man could create. Its possession conferred legitimacy, or at least plausibility, to Louis's claim to be a world leader.

Unfortunately, though, the organ was no substitute for armed men or an easy succession. Far from reclaiming the empire of the East, Louis's heirs frittered away what they had. Edward Gibbon summed up the fate of the Frankish empire with his usual vicious clarity: "The vast body had been inspired and united by the soul of Charlemagne; but the division and degeneracy of his race soon annihilated the Imperial power, which would have rivaled the Caesars of Byzantium, and revenged the indignities of the Christian name." This was the true dawn of the Dark Ages —as Gibbon went on to write: "Every peasant was a soldier, and every village a fortification; each wood or valley was a scene of murder and rapine."

Amid the darkness, the disorder, the collapse of civil rule, the church remained. It had discarded what was irrelevant to its purpose and its vision. As Augustine himself wrote: "Whatever knowledge man has acquired outside the Holy Writ, if it be harmful it is there condemned; if it be wholesome, it is there contained." But the church seized hold of what might serve its ends. It took a campaign from within, but the church retained for itself the organ. Its music became the symbol not of temporal power but of spiritual transcendence, and from the tenth century to the fifteenth the organ belonged almost solely to God.

The transformation of the organ into an engine of divine worship did not happen without a struggle. Augustine's objection still held, at least for some. As late as 1166, with organs in fairly common use within the liturgy, Saint Aelred, abbot at Rievaulx in Yorkshire, could write: "What use, pray is this terrifying blast from the bellows that is better suited to imitate the noise of thunder than the sweetness of the human voice.... It is as though the crowd had assembled, not in a place of worship, but in a theater, not to pray, but to witness a spectacle." But for most, the organ served to please, and could be justified as a symbol of religious concord. As Bishop Baldric of Dol wrote in the twelfth century in a letter to the citizens of a town whose abbey church held an organ: "It encourages me to reflect that, just as diverse pipes, of differing weight and size, sound together in a single melody as a result of the air in them, so men should think the same thoughts, and inspired by the holy spirit, unite in a single purpose.... For ourselves, we speak categorically—because organs are a good thing, if we regard them as mysteries and derive from them a spiritual harmony; it is this harmony that the Moderator of all things has instilled in us."

About 1,700 years separate Baldric from Pythagoras, but it is possible to hear the echo of the Greek in the words of that lord of the church—an echo, though, not the original tone. A central Greek goal had been clarity—as Philolaus wrote: "For the nature of Number is the cause of recognition, able to give guidance and teaching to every man in what is puzzling and unknown." Baldric saw mystery, holy mystery in a machine, and drew from its musical harmony a sense of the nature of divine creation. The Greeks' Number, over the centuries, gave way to heaven.

But the organ could play for both. The common theme, the persistent role of the organ, was as a symbol of power. Power over nature as we would see it now, power to imitate nature, technological power to make distant emperors tremble—the particular message varies over time, but the underlying meaning remains.

And even more consistently, the organ served as an instrument with which one could daily explore the link that harmony builds between a perfection we may hear and a perfection that was believed to exist, the perfect harmony of nature, the perfect harmony of the divine.

Baldric was so inspired by its music that he asked: "Are we not organs of the Holy Spirit?" By the twelfth century the church was secure in its sense that evidence of perfection, the experience of beauty, the knowledge of the laws of the four sciences—arithmetic, geometry, astronomy, and music—would all bolster faith. (Those disciplines, called the quadrivium, formed half of the medieval curriculum for learned men.) God could be revealed within the numbers that regulate the lengths of organ pipes; could be heard in the melodies those pipes could sound. The organ was a tool, the church's tool.

And, finally, to take the allegory of Catherine's wedding seriously, it was the symbol, as Baldric suggested, of the harmony that could exist between God and humankind. It accompanies the mystical union between a living woman, Catherine, and the Son of God. The possibility exists, realized by the saints, to achieve that unity of existence that binds a human being to the divine. Harmony can extend from the material world of daily life—that pleasant room, the city behind it—to the eternal world of heaven, of the Madonna, and Jesus, and the angel playing the tune.

And yet, there is another message, unintended by the painter, discernible within this image. With the acceptance of the organ by the church, after whatever struggle, medieval craftsmen could study the instrument, refine it. By the twelfth century, when Baldric wrote, the small, portable instruments were beginning to appear in some numbers, enabling their makers to take God's music with them, instead of fixing it to the walls of a church. In the thirteenth century small, delicate portatives become more common, compact enough for minstrels to carry. In 1450, or thereabouts, a physician cum astrologist and instrument builder named Henri Arnaut, from the town of Zwolle in the Netherlands, published the best surviving text on the construction of medieval organs, and included within it the specifications for a portative very like the one Memling painted. In 1467, Memling in Bruges sat a child down at the instrument, adjusted her pose, placed one hand upon the wooden keys, moved the other out of sight, and began to paint. In his picture, that child became an angel that sits to this day, removed from the flow of time, as angels are, caught forever between one note and the next.

But if the angel was exempt from the confines of time and space, the instrument was not. A portative could be carried from room to room. While Catherine still commanded the music of heaven, the portative, like all the varieties of organ that preceded it, embodied the priorities of its makers. Even Catherine stands in a private chamber, not a church—and it is her wedding day; the portative could play nuptial songs as well as hymns, for any-one, not solely for priests and congregations. In the High Middle Ages European men of letters read the Greeks again, after a lapse of centuries. New music was being written and played, to be heard by new audiences. That music survives, and so does a version of Catherine's portative. In its melodies we may hear the first sounds of what we now call a revolution.

In the end, Augustine and Aelred were right: the organ is a marvel, a distraction, if what one seeks is the uninterrupted con-templation of God. The organ, as Aelred clearly understood, was a device calculated to feed human vanity. From its origins through to the end of the Middle Ages, it was put to a series of uses—first engineering test bed, then circus toy, tool of imperial diplomacy, finally, as an adjunct to the celebration of the mass—but it always retained its essential character. It was a machine that embodied the attempt to apply human reason to the discov-ery, the mastery, of nature. It was present when a very small community of men created the first glimmerings of the notion that all of nature might be encompassed by a single theory. It survived, barely, to reemerge at a time when belief in the intel-ligibility of nature was about to assault the core of mystery upon which the rock of faith stood. Aelred and Augustine, saints both, had sensed it: danger lurked in the sweet sounds of music, in the intricate machines that human hands would build; in the ideas that guided hands and danced in time to music that some had said only Pythagoras could hear.

Chapter 2

ALL IS

FOUNDED IN

PERFECTION

*When the sun falls just right on the pyramid of
rock* that stands on tidal flats at the boundary between Brittany
and Normandy, the traveler can see what seems to be an arrow of
blinding gold pointing upward from the highest steeple on Mont-
Saint-Michel. The gilded statue of the archangel Michael, pro-
tector of the Normans, sometime patron saint of France, holds a
sword upright; Mont-Saint-Michel, the archangel's shrine, was a
soldier's church. At its peak, its abbots were powers in the land,
chief servants of Normandy's rulers—William the Conquerer
feasted there, and his heirs came too. One day in the year 1158,
Abbot Robert of Torigny greeted Henry II, king of England and
Anjou, and his wife, Eleanor of Aquitaine. The royal party
stopped at the abbey, celebrated mass, and dined that night in
the refectory. In the mind's eye, it is possible to see dusk beginning to fall, and the monks preparing to perform the next of
the offices of their daily worship.

With the setting of the sun the royal party enters the church
atop the rock for vespers, the office or service within the daily
cycle of prayer with which the celebrations of feasts and Sundays
begins. The church is relatively small, about 230 feet long, timber roofed, with towers added to the original structure by Abbot
Robert himself. The space fills with the king, queen, and their
barons, the abbot and his monks. They pray silently. Then the
choir sounds, split in two to alternate during the responsive
prayers: "Make haste, O God, to deliver me; make haste to help
me, O Lord."

Five psalms follow, sung back and forth across the church. A
monk reads "the chapter"—a short text from the Bible, perhaps

chosen to recognize the special relationship between the royal guests and the abbot, godfather to one of Henry and Eleanor's children. After the reading, the choir launches into the elaborate chant called a responsory. The ornate melodies of the responsory provide a test of their skill, a chance to display their mastery of the subtle and precise requirements of the ancient chants of St. Gregory. Another hymn is sung, then a short prayer leads to the Magnificat, Mary's hymn of thanksgiving to the Lord, which forms the musical center of the service, sung responsively once more, choir answering choir. The service ends with prayers and a blessing. As darkness falls over the mount and abbey, we might hear with the king, the queen, and the barons of their court listen to the choir chanting the final prayer of the service: "Benedicamus Domino" ("Let us bless the Lord")—and then its response, "Deo Gratias" ("Thanks be to God"). Then silence, with the last echoes dying against the stones of the church.

To those present, vespers would have come and gone as it did every night, varying with the special demands of feasts and seasonal changes in the liturgy, the cycle of prayer turning constantly over the days and years. The next night the abbot would have had other guests—though few on record were so notable as Henry and his formidable Eleanor—the monks would chant again, and their devotions would continue, around and around, stable and eternal. Yet, within a very few years, changes would sweep through the world all those present that evening at Mont-Saint-Michel would have known. At the core of the services of the church, at the core of the way medieval thinkers understood their surroundings, lay the idea that all experience served to expose the design of God, to demonstrate the truth of what God had already revealed in his creation. But the outcome of the effort to read God's hand was to undermine the foundations of that belief just as surely as Mont-Saint-Michel's towers bore down on their supports. Even Robert's towers would fall, weakened at the base as their weight wore away the foundations.

The transformation occurred throughout the world of thought. But between the tenth century and the fourteenth, medieval men and women created a whole new apparatus of music that typifies the change. Basic advances in the theory of harmony and of rhythm produced by the end of the Middle Ages a body of work unlike any that preceded it. And on the technological side, the pinnacle of medieval musical invention was the persistent improvement (by the church) of the organ until it reached what is close to its final form.

The organ is silent at first; this story begins with choir song unaccompanied by any instruments, as King Henry heard it at Mont-Saint-Michel, and as the church first favored its music. The music of the liturgy of the Roman Catholic church was one of the church's most important possessions, built and broadcast at the command of one of the most remarkable men in the history of Western civilization, Gregory the Great, bishop of Rome, pope, and saint.

Gregory was a Roman, one of the old nobility, born about 540. He was wealthy, a prefect of the city, but renounced his estate, establishing a Benedictine monastery instead. From there he was noticed by the church hierarchy, appointed a deacon around 578, and dispatched to the imperial capital in Byzantium as the papal ambassador. He returned to his monastery as abbot in the middle 580s, and then, in 590, was elevated to the papacy.

Gregory and his church faced adversaries at every turn. Italy endured the constant menace of Lombard invasions. The Moors threatened Roman Africa. Spain was crumbling into anarchy as the Visigothic kingdom eroded from within. Paganism had revived in England after the Saxon invasion ended Roman rule there. In response, Gregory asserted twofold authority: that of the Catholic church and of his own, as the successor to Saint Peter. He congratulated the Visigothic king Richard on his conversion to Catholicism. He sent Saint Augustine (not the philosopher-bishop who wrote against music two centuries before) to England with thirty-nine Benedictine companions to retake that kingdom for the faith in the year 596. He set the church's finances on a (temporarily) sound footing. And in his own person he established the principle that the bishop of Rome was the supreme authority in Christendom, with jurisdiction not only within his own diocese but over the entire church.

In the midst all this political maneuvering, Gregory found time for what seems at first glance an odd distraction: reorganizing the *schola cantorum*, the papal choir and singing school founded 150 years before to foster the performance of religious song. Under Gregory, the schola collected and codified the melodies used to sing the worship of the Catholic faith: the body of music later called Gregorian chant. Legend has it that Gregory himself was a brilliant composer who was responsible for at least some of the chants of the Roman liturgy, but there is little evidence that he had any musical skill at all. Rather, his was a gift for organization. Before his codification of the chant, every major territorial subdivision of the church had its own form of chant, its own order of prayer, its own traditional forms. National churches are vulnerable, prey to the caprice of history—or at least potentially disrespectful of central authority. Gregory's chief goal was to bind together by a common practice all the elements of the Christian community, transcending any political boundary. The music of the church was ready to hand—the perfect tool for the job.

Thus, when Augustine and his missionaries came to England, they brought the Roman liturgy with them. By the year 630, Canterbury already possessed its own Gregorian chant school, and within a century the older, Celtic liturgy was virtually extinct. Singers from Rome brought their rite to King Pepin in Paris, and Charlemagne supported the new form. Ancient, traditional religious music proved tenacious, even in Italy itself, but over time Gregorian chant became standard throughout the territories of the Western church.

The key to this success was the music itself, so well adapted to its purpose that it has remained in continuous use to the present day. At its core, Gregorian chant is the soul of simplicity. It is monophonic—there is only a single line of melody, sung by choirs of male voices. The chant melodies encompass a narrow range—rarely more than an octave, often as little as a fifth or a sixth. In most instances they are made up of stock figures, mosaics of melody, that can be reassembled to build new chants. They have no fixed rhythm, movement in the music following the meter of the words of the prayer being sung.

It is hard now to hear clearly what the ancient chants express. In its time, chant was a device to transmit as effectively as possible a universal message, intelligible to the faithful throughout Christendom. Since the Middle Ages, though, audiences have listened to music with a different sense of what it says, what kind of experience it communicates. The critic and sometime composer E. T. A. Hoffmann wrote of Beethoven's Fifth Symphony: "Beethoven's music opens the floodgates of fear, of terror, of horror, of pain." Leonard Bernstein, responding to Stravinsky's *Oedipus Rex*, could say: "The main melody of Jocasta's aria is like a hootchy cootchy dance," and would later cheer that the piece demonstrated "the amazing power of the subconscious at work," creating a "single new metaphor at the abstract level of pity and power, a single manifestation of that primal antithesis, love and death." Bernstein loved Stravinsky, and Hoffmann, earlier than most, recognized Beethoven's extraordinary romantic power. Their response to the music traced particular meanings, associations, and emotional content within the grand themes—fear or terror, love and death—that the composers sought to explore. They heard, we hear, music created in a context of individual expression, of the composer's feelings or particular ideas.

But if we try to listen to Gregorian chant in that same way, we miss what the men who made it heard. While we experience music, art, as particular, the voice of one heart and mind speaking to another, Pope Gregory's chant was formal, elevated, the musical expression of the fixed and certain connnection between God and man. The music expressed universal reality, not just one person's—the composer's or the listener's—experience of it. We do not hear the sounds of science in Gregorian chant—but in their terms, the members of the medieval church did.

That is, the church fathers sought to tame Pythagoras, replacing his cosmology with their own. The certainty music seemed to offer attracted them. Boethius (c. 480–524), the most influential philosopher of music in the Middle Ages, set music apart from the other disciplines of the quadrivium: "The others [arithmetic, geometry, and astronomy] are concerned with the pursuit of truth, but music is related not only to speculation but to morality as well. . . . The soul of the universe is united by musical concord." For Boethius music expressed essential characteristics of human experience—and it did so absolutely: its truths would be understood in the same way by all listeners. Boethius claimed that "discipline has no more open pathway to the mind than through the ear. When by this path rhythms and modes have reached the mind, it is evident that they also affect it and conform it to their nature. . . . And we should above all bear in mind that if in such a matter a series of very slight changes is made [in the music], a fresh change will not be felt, but later will create a great difference and will pass through the sense of hearing into the mind."

But for Boethius and his followers, "science," music, provided more than a description of truths of nature: it was in fact an example of those truths, the thing itself. Perfection in sound did not simply resemble the harmonious order of the heavens, or describe the concord between man and man; it *was* an instance of all those harmonies. Cassiodorus (c. 480–c. 585) argued:

The discipline of music is diffused through all the actions of our life. First, it is found that if we perform the commandments of the Creator and with pure minds obey the rules he has laid down, every word we speak, every pulsation of our veins, is related by musical rhythms to the powers of harmony. Music indeed is the knowledge of apt modulation. If we live virtuously, we are constantly proved to be under its discipline, but when we commit injustice we are without music. The heavens and the earth, indeed all things in them which are directed by a higher power, share in this discipline of music, for Pythagoras attests that this universe was founded by and can be governed by music.

Isidore of Seville (c. 560–636), Pope Gregory's contemporary, was the third great early medieval writer on music. He wrote that "without music no discipline can be perfect, for there is nothing without it. . . . But just as this ratio [of musical numbers] appears in the universe from the revolution of the spheres, so in the microcosm it is so inexpressibly potent that the man without its perfection and deprived of harmony does not exist." That's hard to hear today: to modern ears Gregorian chant has a haunting quality. It seems pensive, the sound of men contemplating God. Medieval congregations may have felt that as well, but the chant contained this other realm of meaning too. The melodies heard at every service fixed listeners within the order of nature and of human affairs as God had designed it for the faithful. When Henry and Eleanor heard vespers, they were given not just a picture of God's grace, but the truth of it.

This central importance of music is what led the church to foster the study of music theory and performance, beginning, but not ending, with Gregory's reform of the schola cantorum. And that emphasis led in turn to one of the fundamental inventions of the Middle Ages. Writing in the early seventh century, Isidore defines music as an "art of modulation consisting of tone and song, called music by derivation from the Muses," and that "it was fabled by the poets that the Muses were the daughters of Jove and Memory. Unless sounds are remembered by man, they perish, for they cannot be written down." All Europe, that is, was without a literature of music.

At one level, the lack of notation did not matter to church musicians. The words of prayer were what counted, theologi-

cally, and those words dictated the music, setting the rhythm of
the chant to the rise and fall of the prose texts (which could be
written down). The melodies were transmitted orally from master to student. But the number of chants grew so rapidly as to
strain the memory of even the most devoted student, which led
to the first attempts to represent musical instructions in written
form. The earliest examples came in the eighth century. They
were chant texts with signs, called *neumes*, written above the
words. Some of the oldest, simplest neumes include the /, called
virga, a rod, the sign to sing a higher note, while the sign for a
lower note, originally \, became a •, or *punctum*. Signs could be
combined to indicate that one word should be sung with several
notes, but there was no information about the actual pitch of the
notes, their duration or their rhythm. The neumes could remind
a singer of a melody he already knew, but for a chorister starting
from scratch, they were useless.

The next step was the invention of what are called heightened
neumes, neumes that are written on a staff, either imagined or
actually indicated by one, two, or finally four lines. Any fixed
sense of rhythm and duration was still missing, but now the bare
outline of the melody could be read with the text. Nonetheless,

Heightened neumes on a thirteenth-century manuscript. The markings above the
words provide melodic information, but no cues as to pitch or rhythm.

learning to sing remained a lifelong endeavor. The apparent simplicity of Gregorian chant actually provided for an almost endless variation among particular chants, with the singer being required to memorize, note by note, a prodigious volume of sacred music. Guido of Arezzo, writing in the eleventh century, complained: "What is the most dangerous thing of all, many clerics and monks of the religious order neglect the psalms, the sacred readings, the nocturnal vigils, and other works of piety that arouse us and lead us on to everlasting glory, while they apply themselves with unceasing and most foolish effort to the science of singing which they can never master."

Guido set out to correct the error of their ways, and his innovations have shaped the development of music to our own time. He was under no illusions concerning the importance of his work. He wrote to a friend, "I have brought to you as to many others . . . a grace divinely bestowed on me." This was a man determined to leave a mark on time.

He did so with two great inventions. The first was of a system of musical notation that evolved directly into our own. He used a four-line staff, rather than our current five lines, but the underlying principle was the same: "The sounds, then, are so arranged that each sound however often it may be repeated in a melody, is found always in its own row. . . . Some rows of sounds occur on the lines themselves, others in the intervening intervals or spaces." With the fixed rows marking set notes it became possible for the first time to represent the precise form of any melody.

At the same time, Guido formalized the understanding of the so-called church modes, the scale structures that most separate chant from modern music. The eight modes of medieval music differ from modern scales in that each has a different order of whole-note and half-note intervals up or down the octave. In effect, medieval musicians used eight different kinds of scales, where we use just two, major and minor. Any given chant could remain in a single mode, shift modes, or even, occasionally, employ two modes at once. Guido's invention of the staff and systematic musical notation eased the dissemination of chants, but sight reading, especially with the complexity of the modes factored in, was a daunting prospect for many.

To help, Guido invented a teaching tool we still use today. From a popular hymn to John the Baptist that had a stepwise rising melody, he took the syllable and the note that began each

Square notation on a four-line staff reveals relative pitches, along with relative durations of long and short notes.

phrase of the song, forming a series of six notes: ut re mi fa sol la. Replace ut with do, and add ti at the end of the chain, and you have our version: do re mi fa sol la ti. Guido used his system as a mnemonic device. He took his rising set of six notes, called a hexachord, and overlapped it with another, still using the ut-re-mi-fa-sol-la names, and then another. Depending where one started in this overlapping, stepwise series of notes, each mode could be found. And if the singer simply remembered the original hymn, he could use the system to pick out melodies he had never heard. "To sing an unknown melody competently as soon as you see it written down, or hearing an unwritten melody, to see quickly how to write it down well," he wrote, "this rule will be of the greatest use to you." Guido claimed that with his methods he could now produce "a perfect singer in the space of one year, or at the most in two," whereas previously, ten years of study yielded "only an imperfect knowledge of singing."

To Guido, his inventions were simply more efficient tools with which to describe to ignorant choristers what the body of religious music already contained. But nothing in the techniques of writing music restricts what can be written. The invention of musical notation brought the creation of music itself into the reach of anyone (with the necessary talent) who could read. Like Gutenberg, who printed his famous Bible with technology that would be used to challenge the church's authority at every turn, Guido's creation opened up vast realms to musicians with interests far beyond the elegant and austere sounds of chanted prayer.

Guido d'Arezzo died around 1050. Already, in his own time, the beginnings of the single most important transformation in Western music had been set in motion: to the pure melodies of chant, musicians had started to add notes above or below the main line. This vertical arrangement of sound—one tone laid atop another—was called polyphony, meaning many voices. Its emergence marks the first step down the road that has led to modern harmony. The vertical harmonies essential to three-chord rock-and-roll and a Beethoven symphony alike descend, ultimately, from medieval polyphony.

The earliest examples of polyphony, however, were extremely simple: chants with a single drone note below the melody, or versions with the melody repeated note for note above or below the main voice at one of the pure Pythagorean intervals—fifth, fourth, or octave. Such early forms of polyphony were known by a word derived from the Greek: *organum*, from the root meaning

A late Renaissance diagram of Guido's ut-re-mi . . . musical instruction system. The scales overlap on the scaffolding of an organ, to show the notes of each of the medieval modes.

instrument, tool, or, in a slightly different form, the organ. The timing may be a coincidence, but the elaboration of this new music proceeded at the same time that the medieval church fostered the development of an instrument, a machine perfectly suited to the performance of such music—both with the same name.

Reenter the organ, though somewhat unpromisingly, for it took some time before the technology of organ construction caught up with the growing musical ambition of those who built the instruments. Initially, large church organs had a very simple job to do: to awe, to overwhelm. One of the most famous early church organs was built in the last years of the millennium at a Benedictine church in the town of Winchester, south of London. Sometime in the 990s, the poetically inclined monk Wulstan wrote a letter in verse describing the instrument. He boasted: "And the melody of the pipes is heard everywhere in the city/and fame goes flying through the whole country/Thy solicitude has dedicated this ornament to the mighty church/And has built it in honor of Peter, the Blessed Turnkey."

By Wulstan's account, that organ was a spectacular machine. It had, he wrote, twenty-six bellows "which 70 strong men stir up/moving their arms and dripping with much sweat," driving air up into a curved windchest "which alone supports 400 muses [pipes] in order." This enormous organ was controlled by a set of forty sliding rods, which opened or closed a rank of ten pipes "as the fixed song of various notes requires. Two brothers of harmonious spirit sit" at the organ, "each the master of his own manual." When things were working as they ought, "here some tongues run, there some return/giving each note its proper value."

Wulstan's letter has something of the air of poetic license to it—the numbers in particular sound more like an assertion of size and grandeur than an accurate count. But it is clear nonetheless that the church at Winchester held what must have been a spectacular example of the technology of the day. The emphasis throughout is on the outsize scale of the device. The twenty-six bellows, probably derived from bellows used in metal forging, imply a huge demand for air—not surprising, given that each note was produced by the massed sound of ten pipes blowing at once. The organ was clearly difficult to play. To play a note the organist would have to pull on a long rod and then push it back into place to cut off the flow of air to the pipes—a one-hand–one-sound motion, compared with the one-finger–one-sound action of a modern keyboard. Such a cumbersome arrangement probably accounts for the need for "two brothers of harmonious spirit" to play an instrument that had just forty separate notes.

Unfortunately, all that labor probably did not produce much in the way of sweet music. As Wulstan describes it, the instrument was optimized for volume. The multitude of bellows could generate lots of air but not a stable, steady pressure. The result, in any organ, would be wild variations in the pitch its pipes would sound. The Winchester organ must have made a mighty noise—but Wulstan himself suggests that it was more awesome than beautiful:

> *Like Thunder, the strident voice assails the ear*
> *Shutting out all other sounds than its own;*
> *Such are its reverberations, echoing here and there,*
> *That each man lifts his hands to stop his ears,*
> *Unable as he draws near to tolerate the roaring*
> *Of so many different and noisy combinations.*

In its cacophony, the Winchester organ would have been typical. The tenth-century organs Saint Dunstan built at Malmesbury used "bellows . . . tortured into belching forth blasts of air." Such instruments were curiosities, marvels, freaks of manufacture. The organ's inhuman roar, with its "thousand breaths," its "strident voice," could overmatch those who required majesty of their church. The giant organ, at least, was not integral to the church service itself; it seems to have been used more like a barker's come-on, to lure people in.

Such gimmicks may have been necessary. The end of the millennium was a bad time for most in Europe, and it was a particularly disastrous era for the prestige of the church. Barbarians threatened Europe from each point of the compass—Vikings from the north and west, Magyars on the east, Saracens on the south. The poor had much to fear from petty local wars which sputtered constantly. Gibbon, writing of the Normans who conquered Apulia in Italy, noted that "every object of desire, a horse, a woman, a garden, tempted and gratified the rapaciousness of the strangers, and the avarice of their chiefs was only colored by the more specious names of ambition and glory." And for every prince of the sword turned thief, there was, it seemed, a prince of the church to blast the spiritual lives of the faithful. Pope John XII, bastard son and grandson of popes, was deposed in 963, accused of a variety of spectacular crimes: castrating a cardinal, drinking to the devil, running a brothel out of his palace. The lives of the popes could be nasty, brutish, and often abruptly shortened. Six primates were assassinated within a century. The church as a whole seemed to mock itself.

But by Wulstan's time, signs of change had begun to appear. The Winchester organ, in fact, was completed at a time of rapid change in medieval ways of life as Europe finally began to catch up to other, more advanced civilizations. For all the catastrophes of the years leading up to the millennium, in the eleventh century new ideas emerged rapidly, spreading across Europe. The horse collar, probably imported from Asia, was replacing the yoke, which choked a horse who strained too hard against a load. The collar multiplied a horse's power three or four fold—thus speeding the slow process of freeing human beings from their own physical limits. The adoption of a three-way rotation of crops increased yields 50 percent over the old two-field system. Mechanical inventions appeared: water power (also long familiar to the Chinese) was in common use by the eleventh century, with gear mechanisms showing up shortly thereafter. Cathedral schools proliferated, offering large numbers of priests and monks a more rigorous education than had been broadly available for centuries. The European climate itself was benign, warmer on average than modern climates. Growing seasons were longer, and the first age of maritime discovery, the Norse voyages west, led to a probable landfall on North America and the establishment of a European outpost (complete with its own bishop) as far west as Greenland.

Against this background, men of learning finally began to reclaim skills that had been lost with the breakup of the Roman Empire, skills essential for the construction of complex machines. The basic failings of the early church organ were that it sounded terrifying and was hard to play. To create a truly functional musical instrument, medieval organ builders had to solve three central problems: controlling the flow of wind into the organ to maintain stable pitches; creating an "interface"—the keyboard—that an organist could manipulate easily; and coming up with a system of pipes and pipe tuning that could sound a joyful noise. The innovations used to resolve these difficulties produced the organ in the essential form it retains to this day.

The organ sound begins with its flow of air under pressure. If the organ's air pressure changes substantially, the sound of any pipe will alter: roaring, fluttering, wandering all around whatever note the pipe is supposed to voice. Instead of a hydraulic regulator, medieval organs like the Winchester instrument took air into the wind chest directly from the bellows, which moved enough air to produce as loud a note as desired. The drawback, though, as the organ hater Aelred noted, was that early bellows could produce enough noise of their own to distract the listener.

The ultimate solution was a system invented in 1762 by Alexander Cumming, a clockmaker, who connected a feeder bellows to a second folding chamber that acted as the reservoir. Clean-sounding organs appeared much earlier, however. Medieval organ builders hit upon a simpler compromise, improving the action of the bellows to make it easier for the blowers to control the flow of air. The earliest versions were little more than bags of animal hide—ram skin, as one eleventh-century theorist prescribed, or even elephant leather. By the 1100s, large organs could be fitted with bellows laid flat on the floor, pumped by blowers using their whole body weight for the downstroke. Ultimately, organs got their wind from arrays of wedge-shaped bellows upon which the organist's assistants performed a kind of rhythmic dance.

Such bellows appear in drawings of the pump room of the

The pump room at Halberstadt Cathedral during an organ performance.

Halberstadt Cathedral organ. Maintaining a steady, constant beat on the bellows throughout a performance would have required both skill and stamina, but the task seems no more difficult than maintaining a constant stroke among the eight rowers of a racing shell. The results must have been at least tolerable at Halberstadt, for the organ there made its way into the regular cycle of worship, performing on the great feast days of the Catholic church. The solution was not as elegant as Ktesibios's subtle hydraulic engineering, but it served the need of the time well enough: it provided a controllable source of wind that could achieve quite high air pressures, capable of producing enough sound within the pipes of an organ to fill the cathedral.

Before the organ could take its place in the church's service, though, it had to be reasonably easy to play. Volume and stable pitches help, but to make music a performer has to be able to control the instrument—and here the primitive quality of early medieval instruments is striking, compared to their ancient predecessors. The Greek organ had a keyboard; Winchester's did not, using instead pieces of wood that slid forward or back to open the foot of a pipe to a flow of air from the wind chest. The first evidence of improvement over the clumsy sliders of the older instruments came in the eleventh century. An anonymous author proposed an organ design that connected the slider mechanism to a large square key mounted on a spring-loaded pivot—essentially the same solution the Greeks had found. On large organs, it could still take one hand to depress the oversize keys, but the springs that pulled the keys back into the "off" position freed the organist to hit notes in rapid succession. By about 1100, the organ had become an instrument capable of increasingly sophisticated performance, and further improvements would follow to create keys with lighter, more playable action.

This is the organ that could have helped spread and develop the organum, polyphony. The habit of sounding more than one pipe with a single slider or key, already present in the Winchester organ, serves as a kind of built-in polyphony, and early uses of the church organ included doubling the vocal line of a chant—creating multivoice music on the spot. Perhaps more important, organs with a keyboard would allow a skilled player to perform two separate lines of music, with each hand generating a distinct part. John Cotton didn't seem to regard polyphony too fondly, for he only briefly touched upon the subject in his book *Of Music*, written around 1100. Nonetheless, he tied the practice directly to techniques of organ performance: "This style of singing is commonly known as 'organum,' insamuch as the human voices . . . imitate the sound of the instrument called the organ." The sarcasm here is subtle—Cotton is suggesting that polyphony was a kind of trick played by singers seeking to imitate a machine. But he still implies that the organ of his day would have sounded pleasing enough to imitate—indirect confirmation of the rapid improvement of organ technology over the previous century.

But if the organ played a part in the creation of polyphony, the new style of music swiftly took off on its own. The simple doubling of chant melodies that characterized many early instances of polyphony gave way rapidly to more elaborate constructions. Composers hit upon the idea of contrary motion—one voice going up in pitch while the other goes down. Some pieces retained the original chant melody (the *cantus firmus*) in one line, usually the lower, while the higher voice sang an intricate vocal decoration that floated above the austere original. Pieces with three and four voices appeared. The possibilities of harmony that resulted from the combination of separate lines represented undiscovered territory, as promising and mysterious as the interior of a newly discovered continent.

Unfortunately, the invention of multipart music forced instrument builders and all medieval musicians to confront one of the basic flaws in the natural history of Number sought by Pythagoras and the church alike. Accommodating the musical intervals, the fifths, fourths, thirds, and so on, created by vertical harmony—two notes sounding together—introduced a problem that was not obvious as long as those intervals occurred only melodically, with one note following another. Careful attention to the construction of a twelve-note scale revealed the impossibility of tuning the instrument so that every one of the Pythagorean consonances would be pure and in harmony with itself and the cosmos—actually building organs exposed a fundamental flaw in the musical description of the universe.

Originally, the construction of an organ's pipes had seemed simple. A tenth-century text laid out the rules just as Pythagoras would have. Start with a pipe of whatever length and call it low C. Divide it into four parts, remove one and you have the pipe for the low F, a perfect fourth, related to the C by the Pythagorean ratio 4:3. Divide the C pipe into three, throw away one part, and the resulting pipe length will sound a G—the fifth above a C, representing the ratio 3:2. Then take the G pipe, divide it into three, add a part to it, and the result is the D below G—fixed by the ratio 4:3, a fourth again. The instructions go on, creating an entire scale that translates the tuning of a monochord into fixed ratios of pipe length.

The ratios work, more or less. Later organ builders came up with a slight corrective factor to make up for the eddying of air as it passes over the lips of the pipes, an effect that lowers the pitch of the pipe slightly. But when an organist tried to play two or more notes together in the polyphonic style, some of the combinations would clash viciously. The flaw was not in the construction of the instrument but in the arithmetic of sound itself.

It is a matter of multiplication. Begin with the octave, defined by the ratio of 2:1, and the major third, defined by the ratio 5:4. In the modern Western scale, three thirds fit into one octave—for example, the octave from middle C to the next C above can be divided into the thirds C to E, E to G-sharp, and A-flat to C. (On modern instruments, G-sharp and A-flat are the same note—but that's exactly where the problem lies.) Tuning organ pipes perfectly, one could find each of those thirds by Pythagoras's method: take the pipe used to create the low C, divide it into five parts, remove one, and there is the E pipe (using the correction for the lip, of course). To build a G-sharp, divide the E into five parts and remove one. To find the high C, divide G-sharp into five parts again, and once more remove one fifth of the length—thus completing the octave.

Unfortunately, while each third thus produced would be perfectly in tune with itself, the octave would set dogs barking. Based on the surgery just completed, the length of pipe, and hence the frequency of the higher C compared to the lower one, will be determined by the calculation $5/4 \times 5/4 \times 5/4$—the ratios of each third multiplied together. The sum yields a ratio of the higher pitch to the lower one of 125/64. But a note one octave above another has been understood since Pythagoras to be exactly two times higher than the original, or 128/64. In practice, those missing 3/64 translate into a hideous, dissonant tone that seems to wobble, or beat against the ear.

Similar catastrophes occur with other attempts to combine perfect Pythagorean ratios into a single harmonious scale on any keyboard instrument. The twelve notes of the Western octave can be derived by leaping up through the cycle of fifths, as Leonard Bernstein demonstrated on the keyboard of a grand piano to his Harvard audience. Begin with a C, go up a fifth to G, then up another fifth to D, D to A, and so on. As the series continues, it picks out all twelve notes of the scale before returning to the original note, seven octaves up from the starting tone. But the two sums—twelve fifths and seven octaves—don't line up perfectly. The cycle of fifths lands slightly higher than the octave, creating a scale with an error of about 2 percent off the perfect ratio of 2:1, a gap named the Pythagorean comma. The essential scaffolding of the twelve-note scale, that is, does not fit the abstract ideal of arithmetical perfection.

In practice such gaps don't always matter. As long as music is purely melody, in fact, the goal of a kind of perfection remains attainable. A choir singing a Gregorian chant can keep each succeeding note in a perfect numerical relationship to the preceding one. Even the simplest kinds of polyphony could survive: purely parallel harmony, with two voices singing the same melody separated by a fifth, or an octave, for example, could march through the tune in lockstep, adjusting each note on the fly to preserve Pythagoras's perfect consonances. But as soon as a composer constructs two lines of music that differ from each other, the incompatibility of the different intervals comes into play. If all the octaves are kept pure and perfect then other intervals must fall out of tune by greater or lesser amounts. To this day every scale we use is "out of tune" in this sense. From the moment that harmony seemed a good idea, the only issue was where to put the dissonances, the imperfections, imposed by the arithmetic of sound.

The first task was to salvage what could be saved of the notion of perfection. Reenter Duke Philip's organ designer Henri Arnaut, the most systematic thinker among medieval organ theorists (at least among those whose writings have survived). Arnaut, working around 1450, came up with one solution balancing the claims of the abstract ideal of perfection against his duke's desire to hear fine polyphony in his chambers. He laid out the length of the pipes to be used, their diameters, even the position of the lips of the pipe openings. Those pipes would sound a scale governed by Arnaut's tuning system, a variant of what was known still as Pythagorean tuning, which emphasized keeping as many fifths pure as possible. Arnaut made every fifth within the scale a perfect consonance except for one, the interval between B and F-sharp. In the ripple effect that occurs with every system of tuning, this decision had consequences for all the other intervals. Only four thirds out of the twelve possible are pure under Arnaut's system, and the sound of the bad fifth is so grating that it was called the wolf, for its wild howl.

That noxious noise was anathema to medieval musicians and was banished from performance, but Arnaut's system and others like it were workable compromises for at least some polyphony. They allowed composers to begin experimenting with the musical effect of harmonic changes: to begin at one note, known as the tonic, and utilize the intervals, the harmonies that develop from that note's relationship to others in the scale—and then to switch, in a maneuver called modulation, to a new tonic note. Arnaut's system represents a kind of bridge between music built out of such harmonic possibilities and earlier music based securely on melody. The large number of pure consonances that remain in Arnaut's pipes permit the playing of melodies that sound very much like those produced by monophonic music, of the sound of a human choir, singing as close to the ideal of perfection as possible. But the presence of the wolf, and of other intervals that are more or less dissonant, impure, limited the harmonic freedom open to the organist. Certain intervals simply sounded bad—but more combinations of notes became available than ever before.

The next step in the push toward harmonic freedom emerged toward the end of the fifteenth century, with the first examples of what was called mean-tone tuning. Where Arnaut had sought to maximize the number of pure fifths, mean-tone tunings sought to increase the number of playable keys. The solution was to spread the dissonance out over a larger number of intervals rather than cramming all the impurity onto a single fifth. The result was to create more of what sound like jazzy blues chords, slightly off-pure, while reducing the number of intervals, particularly thirds, that sound completely unacceptable, jarringly out of tune. There were a variety of different mean-tone schemes, many still preserved in old organs, which would divide the error over four notes out of the scale or five or more. Some intervals, and hence keys, were still sufficiently out of tune as to be unusable, and the wolf fifth still lurked, usually in the interval between G-sharp and D-sharp. But mean-tone systems permitted composers to move through an increasingly large number of different harmonic relationships—creating the modern system of keys, based on scales of similar structure with different starting notes, as opposed to the old modes with their varying scales. Mean-tone tunings allowed relatively easy modulation between certain keys, but not all.

Such systems flourished into the eighteenth century, but all the while, the search for the perfect scale continued. One idea that seemed promising used innovations in instrument design rather than any further experiments in the mathematics of sound. There was an organ at Bucksburg in Germany, built around 1615, that used fourteen notes per octave rather than the usual twelve. On pianos or conventional organs, each black key sounds the note between two white ones—a single key plays both G-sharp and A-flat. The Bucksburg organ split that key between G and A into two and did the same for the key between D and E. By splitting the keys, it became possible to eliminate some of the mean-tone dissonance. The octave from C to C, for example, could now be built out of three perfect thirds, for example, and yet remain in tune: it would rise from C to E, from E to G-sharp, and then from A-flat to C, with the gap between G-sharp and A-flat taking up the missing rise in pitch needed to perfect the 2:1 ratio of the higher C to the lower one.

The idea of subdividing the octave attracted the attention of serious scientists as well as instrument builders. Most famously, Isaac Newton proposed at least two different scales, neither of which seems ever to have been incorporated into working instruments. Later still, attempts to improve on less daring efforts to correct the arithmetical "errors" inherent in the twelve-note scale led instrument builders to wild excess. Henry Ward Poole and Joseph Abbey built two of what they called "enharmonic organs" used in Boston during the 1860s that divided the octave into 36 tones. With such instruments, skilled performers could construct perfect consonances for virtually any set of intervals. Other visionaries attempted similar schemes, planning keyboards that look like elaborate typewriters to control octaves divided into fifty or more notes, but few were built, and fewer loved.

Such schemes were pushed to the fringe early on. Even the simple split key ideas explored in the sixteenth and seventeenth centuries never attracted many imitators, in part because of the difficulties they created for performers, but more because such notions sought to solve the wrong problem. Split keys represent the last gasp of the medieval and the Greek ideal: they are designed, as much as possible, to preserve the perfection of pure consonances. But by the eighteenth century, the pressure of mu-

A sixteenth-century "enharmonic" harpsichord with multiple split keys within each octave.

sical innovation had swept aside that ideal in pursuit of a different one, that of complete harmonic freedom.

The greatest exponent and symbol of the new musical ambition was Johann Sebastian Bach, who claimed the right to compose keyboard music in every key, to move wherever he wished across the twelve notes of the octave. The older tunings, with their howling wolves and their forbidden notes, were unacceptable. Instead he exploited (though he did not invent) an alternative: rather than lumping the inconsistencies between the intervals onto one note or a few, the baroque *well-tempered* scale distributed the imperfections of musical arithmetic across all the notes of the scale, preserving only the octave in its required absolute consonance of 2:1.

That scale was still not quite the final word. The scale used in modern instruments today employs a system called equal tempering. To create a scale in equal temperament, the octave is divided into twelve equal intervals. Every note of the scale rises above the previous one by the exactly the same amount: the precise ratio of the twelfth root of 2:1. The twelfth root of 2 is the number that when multiplied by itself twelve times equals 2. Twelve steps up the scale, then, yields a note twice as high as the one the scale started with, thus preserving the perfect ratio of the octave of 2:1. All the keys then preserve the same characteristics of consonance and dissonance, and the composer has complete freedom of movement up and down the keyboard. By contrast, the well-tempered tunings Bach would have used spread the inconsistencies out in slightly irregular increments across the octave, preserving some of the variety of emotional color for different keys. A keyboard instrument tuned to sound best in C major, for example, would produce considerable dissonance when played in F minor—a key baroque musicians used to express great sorrow. Nonetheless, well tempering smoothed out the scale sufficiently to permit Bach to compose music in all keys, at will. (And actually, in practice equal tempering still preserves some of the variation between keys. Most piano tuners, for example, do not tune their instruments to absolutely exact numerical specifications—they and performers complain that such arithmetically accurate tunings sound dead and dull. So instead, they tune by ear, and there is, almost always, some texture, some ebb and flow in the intervals they actually impart into their instruments and to the music such instruments produce.)

Moreover, equal tempering itself is as at least as old as meantone tuning. Fretted instruments—guitars, lutes, viols, and others—employ equal divisions between pitches. Composers writing for such instruments ranged freely among the different keys long before Bach began to reshape the keyboard repertoire. But, nonetheless, Bach's creation of a body of music that compelled the use of well-tempered keyboard instruments remains a crucial step, the last culminating blow to the concept of musical perfection.

Book one of *The Well-tempered Clavier* was originally written as a set of lessons for Bach's eldest son; it can now be seen as a virtuoso display, as if Bach were showing his colleagues the potential inherent in harmonic freedom, a palette of harmonic effects to be exploited to the full by nineteenth-century composers. Each of the keys now fully available for use had traditionally evoked its own qualities of feeling. D major, for example, elicited pomp and grandeur. (The martial quality of the key of D, according to MIT's Lowell Lindgren, also derives from the fact that D is the fundamental pitch of one of the most popular sizes of trumpet in use in the seventeenth and eighteenth centuries. Until the addition of valves early in the nineteenth century which enabled trumpets to play in any key, many popular flourishes and fanfares were played in the key of D.)

Other keys could be "very melancholy" or "harrowing" or "full of dreamy expression," and so on. But with well tempering, and even more with equal temperament, this correlation of key or mode and meaning eroded, then vanished. The sound and feel of music based on the equal-tempered scale is qualitatively different from all previous harmonic compromises. The octaves, as required, retain their perfect consonance in equal temperament. But every other interval, from the smallest to the largest, is impure, out of tune, and every piece of music we listen to today is built upon this tower of discord. Just as Pythagoras's followers believed that they could not hear the music of the spheres, having been surrounded since birth by its constant drone, today we simply ignore the dissonance, and focus instead on the flights of harmonic fancy our scale permits.

To medieval ears, though, even those sound-drunk with the new possibilites of polyphony, such an exchange was unacceptable—the lure of perfection, to be preserved as much as possible, remained strong. Better, as Arnaut recognized by the lights of his time, to harbor a wolf and some bad thirds to save the rest of his scale than to accept what today we do not notice but yet remains audible: that slight quality of dissonance in every move from chord to chord. But medieval polyphony launched the process that has led to this conclusion. As the experiments with harmony progressed, the music being made moved ever further from the idealized worldview contained in the traditions of church music. The years between the height of the Middle Ages, around 1200, and the end of the period, around 1450, saw the development of musical apparatus, tools with which to build what was recognized at the time as an *ars nova* (a new art). That music could no longer support the old connection between music and the Pythagorean cosmology of the church. Instead, its compositions increasingly revealed a sensibility recognizably closer to our own than to that of the chant from which it originally descended.

There were songs that recalled the triumphs of the hunt—one, performed in three-part harmony, captures the moment of the kill: "Ho, or tout coi/ho, je les voi/ ho, jetes, jetes, ou vous les perdes" ("Ho, quiet there/ ho, I see them/ ho, throw, throw, or you lose them"). Love, loss, and desire obsessed the poets and their audiences. Guillaume de Machaut, born around 1300, was the greatest of the French composers in the ars nova style. He exclaimed that "when I've been to see my lady, I feel neither pain nor sorrow, by my soul. . . . The memory of her beauty and great charm excites and inflames me night and day with passion." And, in less happy days: "Alas! continually without respite I adore/your fair sweet face,/but I can find in it no kindness or love/Only a hostile look." Courtly love had its place, and so did flattery, along with a growing political edge. Paolo da Firenze gloated over the defeat of Pisa by Florence in 1405 with a madrigal for three voices, exhorting his city to "Rejoice, rejoice, Florence, since you are so great."

Such a boast completes the leap in spirit and in feeling from the world contained in the opening cry of vespers: "Make haste, O God, and deliver me." Gregorian chant was born of mostly anonymous, collective genius, the expression of what was recognized as a universal relationship between humankind and the cosmos. Some chants, it is true, like those composed by the mystic, passionate Hildegard von Bingen, reveled in the personal experience of religious emotion. But this music, this new art, was not only singular, individual, the creation of particular minds and voices. It also sought to produce a whole new radical experience of sound. The movement gained its name from Philippe de Vitry's manifesto *Ars Nova*, written around 1320, while Phillippe's contemporary Jean de Muris weighed in with a treatise called *Ars Nova Musicae*—the new art of music—in which he laid out the principles for the creation of subtle, varied, original music.

The critical discovery de Muris and his allies made was that of time, of the potent force of rhythm and of the techniques needed to control the measurement of time within the compass of a musical composition. Guido of Arezzo's original invention of musical notation recorded melody but contained no symbols that recorded the duration of the notes. There was no need. Gregorian-chant rhythms flow from the text, from the motion of prose, ordinary speech. Time is continuous, undivided, the river of experience that stretches from creation to judgment day. Verse—poetry—expresses time differently, in beats of some duration; experience—the music—becomes divisible rather than continuous. It becomes a quantity, and in its medieval context, it spurred the making of the tools of measurement.

The first step came with the invention of square notation during the twelfth century. Square notation responded to the requirements of increasingly complex polyphonic compositions, in which the interplay between the several parts of the music put a premium on the ability to compare units of time, to manipulate them. The initial form of square notation was simple, offering only two units of time: a plain square marked a long note, and a square with a tail, a straight line down the side, represented the shorter duration. (Square notation developed a considerable variety quite quickly, but that basic distinction remained.) The short note was precisely one third the duration of the long one—suitable for the writing of music in what thinkers of that day considered perfect time.

Perfection again, again derived from the symbolism of Number: perfection in time, in rhythm, derived from the number three, the measure of the Holy Trinity. Jean de Muris recalled the reasoning of his predecessors:

> That all perfection is implicit in the ternary number follows from many likely conjectures. For in God, who is most perfect, there is one substance yet three persons. . . . At first in knowledge, are the separate and the concrete; from these under the ternary number the composite is derived. . . . Three attributes in stars and sun— heat, light, splendor; in elements—action, passion, matter; in individuals—generation, corruption, dissolution; in all finite time; beginning, middle, end; in all curable disease—rise, climax, decline. . . . Now since the ternary number is everywhere present in some form or other, it may no longer be doubted that it is perfect.

De Muris went on to say: "All music . . . is founded in perfection, combining in itself number and sound."

The language sounds familiar and old, this business of perfection, numbers, the order of the universe and its recapitulation in music. But de Muris saw himself as a revolutionary of sorts, writing in times ready for the change. The transformation came in the consideration of the number two. De Muris laid out a series of operations for combining rhythms based on three beats with those based on two-beat time. Crucially, he and his contemporaries invented the system of French ars nova notation, a method for constructing and elaborating combinations of inconsistent rhythms, imperfect rhythms as they called them. The first and simplest innovation was simply that of writing music using two colors. Black notes retained their older meanings, dividing time in threes—three of the shorter notes, called semibreves, adding up to one breve; three breves adding up to one long. Red notes, though, counted in twos. Commonly, one red note would be two-thirds the duration of the length of a black note of the same shape. The two-color notation created an extraordinary compression of information: a single symbol contained instructions about the pitch of a note, its duration, and the meter, or beat, of the music at that point in the piece. The composer could change all three with every note, on every line of a multipart, polyphonic piece. The range of possible patterns and combinations was end-less—and this realm of possibility was created by the construction of a system that could represent musical information with both sufficient precision and flexibility to allow a composer free rein.

Notation, that is, served as a kind of computer or calculating engine, a tool with which the ars nova composers could place rhythm against rhythm in a series of experiments on the divisibility of time. The result was the most complex rhythmic music written in Western culture between that time and the development of complicated jazz rhythms composed in this century. By the end of the fourteenth century the effort produced what was called *ars subtilior* (the more subtle art), in which the giddy pleasure of rhythmic invention exploded. The chief goal of such music seems to have been to extract the maximum amount of independence for each of the individual lines of a polyphonic song. Composition and performance became an exercise in the making and solving of rhythmic jigsaw puzzles.

There are two famous examples that represent some of the extremes of this passion for intricate patterns, both by a composer named Baude Cordier. One (actually a relatively simple piece) is a rondeau, called "Tout par compas"—which states: "With a compass was I composed,/properly, as befits a roun-delee,/To sing me more correctly." A perpetual canon, it repeats to infinity, and as a virtuoso display of the control of time across all three parts, the piece was written on concentric, circular staffs. The depiction on the page expresses the central idea within the song, the endless return of sound against sound. As one verse put it: "Three times around my lines you posed,/you can chase me around with glee/if in singing you're true to me."

Cordier's control of the different proportions of time the new notation could record enabled him to produce songs like one titled "Amans ames secretement" ("Lovers, Love Secretly"), with as many as ten different rhythms confined within three voices. His most spectacular example of form echoing function, though, came with the song "Belle, bonne, sage" ("Beautiful [Woman], Good and Wise"). The rhythms in this piece are actually somewhat simpler than those in some of his other music and in the work composed by others at about the same time, but Cordier captured the intellectual exuberance of his time in this love song nonetheless. A song offered with his heart to a lady, the score was written in the shape of a heart. The pieces of the puzzle join seamlessly: music, words, the construction of the score on the page all coincide to deliver the message—or rather messages. It spoke of Cordier's devotion and desire on the one hand, and on the other, it trumpeted Cordier's bravura skill, his ability to construct strikingly individual musical statements.

As he helped launch the musical movement that culminated in such tangled and intricate compositions, Jean de Muris attempted to describe how knowledge evolves, how ideas transform. He wrote: "It is not possible for the mind of one man, unless he have an angelic intellect, to comprehend the whole truth of any sci-

Baude Cordier's musical heart. The gray shades correspond to the red notes.

ence. Perhaps in the course of time there will happen to us what is now happening to the ancients who believed that they held the end of music. . . . For knowledge and opinion move in cycles, turning back on themselves in circles." The term *revolution* has a complicated history and a tangled meaning. Here the concept appears in its original sense, in the notion of experience ever revolving upon itself, of change masking the essential change-lessness of cyclical time. But there is a quality in de Muris here of a disclaimer, some balm to ease the lash of the novelty of his ideas. For the impact of the new art of music as practiced in the fourteenth century was revolutionary in the modern sense, as de Muris seems to have understood. The music sounded unlike that of preceding generations, certainly. But more important, the im-plications of the methods and motives of composers overturning the claims of the ancients represented what we can recognize as a revolution in the attempt to organize human experience.

The divisibility of time that entranced de Muris and his suc-cessors illustrates the change. Number to the Pythagoreans had been an absolute—each of the first four integers, 1, 2, 3, and 4, presented a fundamental property of the universe. But when de Muris laid out rules for the intermingling of two-beat and three-beat rhythms, he articulated a worldview in which Number de-scends into numbers, relative measures of quantity: an absolute becomes a tool, a technique with which to calculate.

In the creation of this new music, the notation that could record time precisely served as a kind of conceptual machine, the intellectual equivalent of the mechanical clocks just beginning to appear within medieval society. Both counted the units of time; both accumulated time into measured quantities—and musical notation offered one more advantage: the chance to manipulate time, to experiment with the relationships to be discovered when units of time can be measured precisely enough. De Muris was a prophet as well as a revolutionary: "What can be sung can also be written down. Moreover, there are many other new things latent in music which will appear altogether plausible to posterity."

Musical invention of any sort responded to this novel imper-ative, these developments from within music of new forms of expression. Arnaut's design for a portable organ was a response to the demands of his day. It would have possessed none of the thunder of the great organ at Winchester—a portative's voice is small and intimate, better suited for a private chamber than an open nave. Arnaut's instrument was a tool for performing the kinds of music his master, the duke of Burgundy, would have required: the intricate and complex songs of the day with which Philip's favored musicians would have entertained the ducal court.

This music of the table and the chamber carries with it a conception of the world vastly changed from that contained within the ancient melodies of chant that de Muris, Arnaut, or his duke would still have heard, echoing amongst the columns of the church. In the time of Gregory and the Frankish kings, there was the Word, and the words of Scripture; there was the holy, eternal Number (three-in-one and one-in-three), and a conception of science, any science, especially the science of music, as the elaboration of truths already known. Experience—of nature, of human affairs, of the daily round of prayer and chant—all this could reaffirm revealed knowledge. Each encounter with the world could give depth and flesh to those truths already known and affirmed by faith, but experience was an illustration, the exercise of reason a demonstration of essential verities. "The power of the mind," Boethius had written, charging those who came after him with the obligation to study music, "should therefore be directed to the purpose of comprehending by science what is inherent by nature."

To follow this line of thought, music becomes inadvertently a kind of laboratory within which to examine in sound the operation of laws that govern not just music, but all of creation. Yet almost paradoxically, a belief in the perfection of a musical description of the world creates the impulse to transform that perfect music. If music leads to a deeper understanding of God's design, then more music, better music, a more detailed elaboration of musical expression, might lead to a still greater apprehension of the nature of the universe. The urge to expand the view, as it were, would have been inescapable, built into the longing to examine more closely the truths that music could expose.

And that urge carries with it danger very like that Augustine had sensed: the risk that experiments in sound would expose a gap between the experience of music made and heard and the underlying vision music was supposed to convey. The very organs of the church revealed the disjunction between the perfection of the universe that music was supposed to embody and the discordant arithmetic of the musical scale. The intellectual tools musicians built widened the gap: the innovations of notation originally developed to speed the training of choristers studying the chant opened up a world of possiblity in the creation of elaborate polyphonic harmony and rhythmic experiments. The subjects treated in the new music made possible by such innovations deepened the gap into an abyss, as cries to God competed for the attention of composers with the need to find a setting for the line "You can chase me around with glee."

The gap between revealed knowledge and experience, that is, was filled by curiosity, prying loose the strictures of the established rules of music, producing unintended, yet irresistible results. The explosion of musical ideas between the time of Henry and Eleanor and the end of the Middle Ages, the time of Arnaut and Duke Philip, hinted at a transformation in the sense of what a science was, what a scientist did. From the study of experience to demonstrate external, eternal truths, music had become a tool of discovery, of innovation. Baude Cordier's musical heart celebrates the possibilities inherent not in nature but in this wonderful new toy, this combinatorial engine of pitch, duration, and rhythm. Will sound fit onto a heart—try it and see! How many rhythms can fit on the head of a pin? How far can we range within the scale to produce an effect that exalts—or terrifies or soothes—the soul? What we cannot find, our successors (so said Jean de Muris) will discover, latent in the possibilities within music.

And if music once had the force of natural law, and music changes, then either the laws must change or something else must give. With the introduction of the idea of discovery, that something else was what failed: what was lost was the sense that all of knowledge and all of experience could be organized according to a preexisting order. The new music produced new facts, new experiences, new sounds, not accounted for by old ideas. But while it is easy to see the change in music, the revolution triumphed not by destroying the older assumption of an underlying order but by co-opting it. Divine order does account for all that human senses perceive. That order is accessible to human reason, but its true nature is obscured by the complexity and detail of the world. The task of experiment thus became, as the Middle Ages turned into the Renaissance, not to confirm the pattern but to discover it, to build it. The ars nova composers, with their experiments in harmony and rhythm, still defined their success by the elegance with which they constructed a finished, patterned whole.

T. S. Eliot, a modern man with a curious affinity for medieval thoughts, wrote in *Burnt Norton:* "Words move, music moves/ Only in time; but that which is only living/Can only die. Words, after speech reach/Into the silence. Only by the form, the pattern/can words or music reach/The stillness." The accidents of experience come and go; the living, dies. But patterns exist beyond time, and form and order can touch Eliot's stillness, the core of knowledge or of being. From de Muris to Eliot, from medieval France to the twentieth century (up to the edge of our own time), the belief in an order underpinning the ephemera of ordinary life has survived, felt deeply enough to be passed almost unnoticed.

The Renaissance was just over the horizon of the imagination when Cordier composed his love songs. The first shots of the scientific revolution were still a century and a half away. Yet in the sustained output of music that experimented in time and sound—in Cordier's happy joke of a heart—can be seen the ground upon which those intellectual upheavals turned. The poets of that time speak for themselves. The old images retained their power: around the year 1423, an anonymous poet in the remote medieval outpost of Cyprus wrote that "the word is woven in the flesh/by the strongest of unions." But one minstrel at the same court spoke for the new, singing that he longed "to see Reason face to face." The echoes from both intermingle, can still be heard.

Chapter 3

NOTHING

IN VAIN

Sometime in the summer of 1674, a tradesman—a draper—from the town of Delft found himself aboard a boat on a small lake. The lake, called Berkelse Mere, had a soft, boggy bottom, and each year, as summer progressed, would become murky. Local explanations for the phenomenon were ad hoc at best; one suggestion the draper may have heard laid the change to heavy dew. It seemed an implausible idea, so the man— Antony van Leeuwenhoek—collected some lake water in a glass bottle for deeper study. The next day he placed a drop of the Berkelse Mere in front of a tiny lens, used a screw mechanism to bring the drop into focus, and began to observe.

He saw a tracery of green lines, coiled into spirals; green globules, bunched up; floating particles; and "very many little animalcules ... some being whitish and transparent, others with green and very glittering little scales; yet others were green in the middle and white at both ends, and some were grey, like ash. And the motion of these tiny creatures in the water was so fast and so random, upwards, downwards and round in all directions, that it was truly wonderful to see." Berkelse Mere grew foul in the summer because the warm water filled with a stew of living things no one had ever seen before. Leeuwenhoek (pronounced Layuvanhook) had discovered the first glimmerings of the science of microbiology.

Using his primitive tool to extend the reach of human senses, Leeuwenhoek and his contemporaries did not simply see more; they saw differently than had their predecessors. Through Leeuwenhoek's lens nature exploded: The creation of artificial aids to a human sense—vision—transformed the act of seeing from a simple survey of the scene before our eyes into a kind of human creation—the active quest for some new view invisible to the unaided gaze. The ultimate outcome was a fundamental shift in what people looked *for*—an intellectual upheaval that after the fact became known as the scientific revolution.

The new tool helped create a new method of finding things out, that is—and the story culminates in the great age of scientific discovery in which Leeuwenhoek played his part. It begins, however, in a novel setting within the medieval landscape, the universities that sprang up in towns, independent of at least some of the bonds of feudal control. Bologna's university came first, founded about 1100, but the University of Paris, established in 1150, was the most renowned center of learning in Europe. In 1241 or thereabouts, a young, ambitious Oxford man named Roger Bacon arrived in Paris to begin teaching in the faculty of arts. In his series of lectures at Paris, he embarked on a line of thought that he would pursue for decades, one that led him to attempt to become one of the first of a new kind of man, the professional scientist. Along the way Bacon almost, but not quite, stumbled across the idea of both the telescope and the microscope. He came as close as he did because of his radical conception of science; he fell just short because he was not quite radical enough (as no one is) to sever himself from his times.

Yet what he did was striking enough. In 1267 Bacon sent to Pope Clement V a manuscript that he titled his *Opus Majus—Great Work.* Vast, ungainly verbose, and repetitive, it still stands as one of the first great manifestos of a scientific ideal. Bacon wrote: "There are two modes of acquiring knowledge, namely, by reasoning and experience. Reasoning draws a conclusion and makes us grant the conclusion, but does not make the conclusion certain, nor does it remove doubt so that the mind may rest on the intuition of truth, unless the mind discovers it by the path of experience." A man may know that fire burns, Bacon said, and then reason that contact with fire will hurt, but unless he places his hand in a fire the lesson will not be truly understood: "But when he has had actual experience of combustion," Bacon concluded, "his mind is made certain and rests in the full light of truth. Therefore reasoning does not suffice, but experience does."

Such a thought is a commonplace now. Michel de Montaigne, writing three hundred years after Bacon, would say what sounds like much the same thing, with an ironic twist: "There is no desire more natural than the desire for knowledge. We try every means that may lead us to it. When reason fails us, we make use of experience." Albert Einstein, more subtly, told the Prussian Academy of Sciences in 1921 that "as far as the propositions of mathematics refer to reality, they are not certain; and as far as they are certain, they do not refer to reality." Experience, empirical evidence, retains its central importance, even in the remotest reaches of modern physical theory.

In the thirteenth century, though, such notions were new, exciting, even dangerous. Bacon was one of the leaders in an intellectual revolution that began as a theological dispute, one which has in changing forms persisted to this day. In Paris, Bacon took part in what became a running intellectual war over what a Christian scholar could study, prompted by the European rediscovery of knowledge it had for centuries forgotten that it had forgotten. Throughout the latter half of the 1100s, Latin translations from the Arabic of Greek classics and more recent Islamic scientific texts became available in increasing numbers. Euclid's *Elements*, *Optics*, and *Catoptrics* were translated; so was Hero's treatise on pneumatics. Adelard of Bath and Robert of Chester translated such Arabic works as Al-Khwarismi's texts on arithmetic, trigonometry, and algebra. Rabbi Ben Ezra, a Spanish Jew, explained the system of Arabic (originally Hindu) numbers, introducing the symbol for zero to European mathematicians. Later in the century more texts appeared—the seminal medical works of Galen and Hippocrates, Ptolemy's *Geography*, and others. And looming above all of these came Aristotle. His writings on natural history and science were translated in a flood tide of labor between 1150 and the first decades of the thirteenth century, presenting the thinkers of Europe with a body of work that reached, seemingly, into every field of human inquiry.

In rapid order, scholars devoured studies of physics, metaphysics, logic, meteorology, biology—the extraordinary corpus of Aristotle's investigations. The sum of the separate texts was a comprehensive system that attempted to account for the entire sensible universe through natural causes. Doctors of theology had long dominated the schools, investigating the relationship of humankind to God through one divine text—that of Scripture. In the new universities, where the faculty of theology existed alongside that of the arts, new masters found in these newly restored old thoughts the key to the other text that God provided to medieval minds: the book of nature.

There was one problem. Aristotle was no Christian. A close reading of his philosophy turned up, for the best thinkers of the day, clear conflicts between Aristotelian ideas and the theology of the church. Most important for the debate in the first half of the thirteenth century, Aristotle posited a world that was eternal, with no beginning and no end. Aristotle speaks of God, but only as a prime mover, the immobile, immaterial agent that set the material world into motion. This God inhabits a universe that simply exists: "neither matter nor form comes into being." These ideas leave no room for a creator, for the spirit of God hovering over the waters, for the active power that could command, "Let there be light."

For the scholars, this was first and foremost a proposition to be refuted. The greatest of them all, Thomas Aquinas, sought to reconcile Aristotelian philosophy with revelation by demonstrating (to his satisfaction, at least) that the act of creation and the existence of the Creator could not be proved logically, and could only be understood through revelation. Others simply regarded Aristotle as wrong on this point, guilty of an error in logic—which one could possibly excuse in a pagan, but that would be heresy to repeat in a Christian century. (Or a Jewish one, for that matter—Moses Maimonides labored to prove the necessity of a creator.) Some—including Roger Bacon—simply denied that Aristotle meant what he clearly wrote. The argument is still going on, of course, in a somewhat different form: Stephen Hawking's theory of time lands him, in the end, in an Aristotle-like position, proposing a universe in which time has neither origin nor end, while more conventional cosmologies point to some initiating event that got our universe on track. Such theories say nothing, of course, of what might have preceded the big bang, but as Augustine warned, there is a special place in hell for those who would ask such questions.

If the argument that raged in the thirteenth century had remained within the university, it might have burned itself out in time, vanishing as have most of the passionately fought battles of medieval learning. But to the church as a whole, above and around its learned doctors, this kind of natural philosophy represented a fundamental threat. Aristotle seemed to reach his ultimate theological conclusions through arguments about the sensible world; he read in the evidence of his senses a story which led him to reject (in advance) Christian claims about the nature of God and the divine presence in human affairs. Even worse, consider the provenance of these new (old) ideas: Islamic sources colored Aristotle with what the church considered heresies over and above what was anathema in the original writings.

The heterodoxy of the new learning was too much to bear. In 1210 the archbishop of Paris banned public lectures on Aristotle's metaphysical and scientific works. Pope Innocent III, through his representative, confirmed the ban in 1215. In 1228, Pope Gregory IX warned the theology faculty at Paris against falling into the trap of confirming theology by philosophy, and reinforced the ban against studying suspect works—which included any writings at all on secular subjects. Faith, not reason, was supposed to lie at the core of Christian learning.

In theory, that should have been that for the suspect texts—and perhaps for the future development of natural science in the West, especially with the reiterations of the prohibition coming as late as 1277. What actually happened sheds some light on one of the great mysteries of the European scientific revolution. It used to be asked why other great cultures—China or India, for instance, or the Islamic empire—missed stumbling onto modern science, but it is the exception that is truly baffling: How did European scientific pioneers manage to nurture new ideas in the face of organized, intelligent, and sustained opposition?

One part of the reason is that the new universities were far more difficult to control than any cathedral school or chapter. Town-gown battles raged even then: student outrages so thoroughly antagonized the citizens of Paris that the university as a whole had to shut down between 1229 and 1231. And whether in session or out, members of the university community retained sufficient independence of the ordinary church and feudal hierarchies to continue to acquire and read the growing numbers of Aristotle translations, free of any serious threats.

Nonetheless, Pope Gregory maintained the embargo against teaching Aristotle at Paris after the university reopened in 1231. But in time, as mortal men must, he died. The fight over the succession occupied the eventual winner, Innocent IV, for several years following Gregory's death in 1241. He took the time to renew Gregory's edict against Aristotle in 1245, but by then he was too late, and the edict was unenforceable. That momentary inattention was all that was needed, officially. The intellectual ambition of the scholars, captured by the irresistible promise of the new learning, did the rest. Sometime during the four years of Clement V's benign neglect, the faculty of arts at Paris reinstituted Aristotelian learning as an official part of the curriculum, and found in Roger Bacon a man who could teach the subject.

The young Doctor Bacon was in his early to middle thirties when he took up his duties in Paris. He was proud of his Oxford education and had mastered the traditional skills of scholastic disputation. At this stage of his career he had yet to focus clearly on natural science, and at first must have made a conventional sort of lecturer, working through the Aristotelian texts set for each course using the traditional sequence of question, dispute, answer, next question. But amid the familiar tropes of medieval instruction, Bacon dropped the first hints of the radical direction his thoughts would take over the decades to come. A student's account of Bacon's classroom debates records one such instance.

Bacon's class was considering the problem of what happens to the vegetative soul of a plant once it has been grafted onto a host trunk. (The word *soul* in this context can be read as essence or purpose—the end to which the plant strives.) The first question was: Can the operation be done, leaving both graft and host alive? Bacon answered yes—he had seen it done, and Aristotle concurred. Experience and authority agreed. The dispute continues through several questions about what happens to the "souls" of each plant. By the end of the debate, Bacon reaches what was then a radical conclusion: the two plants each retain their own souls, as proven by the fact that one can see with one's own eyes that both still produce their own leaves and fruit. Aristotle's doctrine of souls required that any union that produces a single organism leaves that organism with but one soul; what Bacon saw convinced him otherwise.

The language is antique, and Aristotle's notion of a plant's soul is almost unintelligible in contemporary terms, but the argument can be updated to make sense. Bacon made the critical observation that the grafted plant bears two kinds of fruit, not one, and recognized from this fact the fundamental truth that host and graft retain, unmixed, an essential characteristic.

It was only a beginning—Bacon made no general claim for a scientific method yet. In his Paris lectures over the space of about four years (the exact length of his tenure remains unknown), Bacon would only occasionally resort to his own experience to advance his arguments. His style remained mostly that of a conventional thirteenth-century master, reasoning from authority, whether it be the Bible or the texts of the ancients. (Also, it is only fair to note that Aristotle himself encouraged biological observers to use the evidence of their own senses.) But beginning about 1250, following his departure from Paris, Bacon struck out on his own, launching into a study that led him, in the end, to propose a whole new realm of Christian learning, knowledge to be gained by the pursuit of the systematic observation of nature.

In the extraordinary breadth of disciplines Bacon attempted to master, he did his best work in the science of optics and vision. He wrote of his optical inquiries: "It is possible that some other science may be more useful, but no other science has so much sweetness and beauty of utility." His path through the topic follows an impressively logical line, beginning with a study of the eye and of optic nerves, and then coming up with a broad theory of vision. Some parts of this theory are pure fantasy. Bacon claimed that "those with deep-set eyes . . . must see further than those whose eyes are prominent," or that "a man in a well or in some other deep place will be able to see stars by day which he will not see at the surface of the well." More generally, he followed one of the prevailing ideas of the day: that the eye sends out some kind of visual force, or ray, which strikes the object being seen, creating the image the eye and brain perceive.

But at the same time, Bacon could write that with the aid of lenses, "the greatest things may appear exceeding small, and on the contrary; also the most remote objects may appear to be just at hand, and on the contrary." His experiments with spheres of glass or crystal analyzed the effects of refraction, or the bending of light as it moves from one medium, like air, through another— glass, crystal, water—of different density. We now know that light travels at different speeds through different substances. The famous figure for the speed of light—186,000 miles per second—is the velocity light travels within a vacuum. Everything else slows light down. The change in speed as light passes from one medium to another produces the bending effect that can be seen, for example, in a beam of light shining through a swimming pool. In his time, Bacon made use of the phenomenon of refraction, employing spherical balls of crystal or glass to show how different properties of lenses could make an object appear to be nearer or larger than it really is.

Bacon was able to generalize from such observations. He recognized, for instance, that when the sun or moon appears to be abnormally large at rising or setting, something analogous to the refractive magnifying of a lens must be taking place—and he argued persuasively (and correctly) that the denser the atmo-

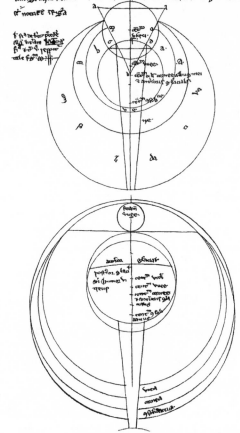

Roger Bacon's diagram of the optics of the eye.

sphere (and hence the greater the refraction) through which one saw the sun or moon, the greater the image-distorting effect must be. And Bacon also understood the practical value of optical instruments. "If one looks at letters or other minute things through the medium of a crystal or glass or other lens put over the letters," Bacon wrote, "he will see the letters much better and they will appear larger to him. . . . Therefore this instrument is useful to old men and to those with feeble sight."

Bacon went further: lenses and mirrors seemed to him wonderful instruments of war, for example, and he suggested that one could use refracted light to bring an image of the sun close over the heads of an attacking army—visions that would ensure victory as "the mind of a man ignorant of the truth could not

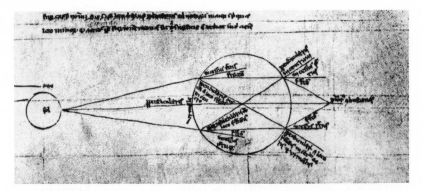

Robert Grosseteste (Bacon's mentor) analyzed the diffraction of light through spherical lenses.

endure them." His love of gadgets led him to propose a wild array of such possible inventions: everything from firecrackers, a Chinese invention whose design he clearly understood, to a flying machine with flapping wings that he was sure had been built, somewhere. (In fact, such a machine, called an ornithopter, finally flew in 1992, seven centuries after Bacon's death.)

There are other discoveries to which Bacon comes tantalizingly close. In his study of magnifiers he almost—but not quite—lays out the design of a simple microscope. And at the conclusion of his section on optics in the *Opus Majus*, he describes what sounds like a telescope—"for we can so shape transparent bodies, and arrange them in such a way with respect to our sight and objects of vision, that . . . from any angle we wish we shall see the object near or at a distance"—and yet never even came close to building one. The problem was not one of laggard practical skills. The technology for at least crude versions of both the microscope and telescope started to become available in Europe by the last decades of the thirteenth century when Italian craftsmen began grinding convex lenses and mounting them in frames to make spectacles. Yet three centuries would pass before anyone actually invented either the telescope or the microscope. Part of the delay was probably due to the need to develop a more systematic and accurate theory of optics on the base laid by the Arabs, Bacon, and his contemporaries. But given Bacon's enthusiasm for cunning inventions, and his obvious passion for optical research, how did he miss?

The answer is, of course, that Bacon did not miss, for he was not aiming in that direction. Though he relished the thought of all of his proposed inventions, Bacon was not a modern man, a genius ahead of his age. Rather, he was an extraordinarily acute, ambitious and energetic thinker, firmly grounded in his time and place. The first task Bacon set his science was to gather facts, to render an unequivocal account of what exists in nature that could be interpreted by the observer. "This science alone," Bacon wrote, "knows how to test perfectly what can be done by nature, what by the effort of art . . . so that all falsity may be removed and the truth alone of art and nature may be retained." The point of such rigor could be found within the fundamental assumption under which Bacon operated. The truths that experience uncovers ensure, Bacon claimed, that "this science next to moral philosophy will present the literal truth of Scripture most effectively, so that through suitable adaptations and similitudes the spiritual sense may be derived."

The understanding of nature, that is, was the means of inquiry, not its end—which was still, ultimately, to know God. There is an aphorism that dates back at least to Bacon's day: "Nature does nothing in vain." What can be found in nature must be. God made it; its existence and its attributes convey some aspect of God's design. By tradition and common practice, men of learning sought God through the study of the Bible and the revealed wisdom of the holy texts. Bacon's inspiration was to recognize that knowledge of God could be found within the book of nature—that all one needed to do was to seek it out, to look at it, to use human vision to measure precisely what the world was made of on the way to the deepest possible understanding of God.

Pursuing that goal enabled Bacon and his contemporaries to see so far and no further. Bacon was a pioneer when he asserted that through experience of the world one may gain true knowledge; but it was knowledge of God he sought in the end, not of nature, not of this world. Experimentation, the discovery of the new, was not the point. Observation was to confirm what was already known to be true, God's revelation. Bacon did not build the instruments to reveal unseen worlds because he was not looking for them. He had no need of such devices, for the world before his eyes was rich enough to serve his purposes. Experience—what one saw and what one felt, the glorious mystery of the divine presence—was the source for Bacon of true knowledge.

Bacon died in 1292, bitter in his last years. The labyrinthine struggles for power in the church trapped him, leading to a period of house arrest in the 1280s. By the end of his life he knew that he had failed to persuade his pope of the excellence of his science for theological inquiry; the idea that God could be sought in nature became a commonplace after him, but his own writings were largely ignored both by Rome and by his brothers in the Franciscan order. Galileo Galilei was born in 1564, into a world of thought that still sought its "sermons in stones" and God's plan in the cosmos. But by the end of his life Galileo would have built a telescope and a microscope—and in the use to which he put his instruments he both drove and was driven by the assault that completely overturned the older view of nature. Like Bacon, Galileo spent his old age under house arrest. In his case, though, it was clearly deserved: what he saw, how he saw it, and the conclusions he drew from what was seen severed the link medieval men had found between the evidence of science and the testimony of faith.

Actually, though Galileo made devastating use of his optical instruments, he was not responsible for inventing either the telescope or the microscope. They emerged not within the academic establishment but from the workshops of craftsmen, the spectacle makers who since Bacon's day had been testing the practical implications of the abstractions of optical theory.

Eyeglasses themselves have a curious history. They existed in the thirteenth century, though no single inventor claims them. The historian G. L'E. Turner has suggested that an unknown craftsman in Pisa may deserve the credit, but there is also very suggestive evidence that both spectacles and dark glasses (made using shaped pieces of pigmented quartz) originated in China and were carried to Europe along the silk route. The word *lens* derives from the Latin *lens*, meaning lentil, and comes to us from the optician's workshop: the first spectacles were made by using a frame to hold two slivers of glass or crysal that bulged outward, convexly, on both sides. The Italian craftsmen who dominated the early trade in eyeglasses called their bean-shaped product *lenti di vetro*—glass lentils. Those early versions served to correct far-sightedness, a failing of vision common with age. Convex lenses bring nearby objects into clear view by bending the light from an object to create an image at the point where the rays of light converge—the focus of the lens. To make a pair of spectacles of the appropriate strength for a given individual was a highly skilled and highly empirical task, for the curvature of the lens had to be ground with care, and adjusted to match the remaining ability of each pair of eyes to bring an image into focus on their own.

The next major innovation came during the middle of the fifteenth century, when some anonymous spectacle maker ground concave lenses to correct myopia, or short-sightedness. A myopic eye refracts too much, bringing the uncorrected image to a focus too soon, in front of the retina. Objects very close to the eye can be seen clearly, but objects at a distance blur. A concave lens refracts light from the object outward, which corrects for the overenthusiastic bending inward of the eye, and allows the short-sighted person to bring the world into focus.

But, following Bacon, and for many of the same reasons, the literate scientific elite ignored such inventions. Discussions of spectacles do not appear in late medieval studies of optics, partly because the convex and concave lenses are far more complicated shapes to analyze than were Bacon's spherical magnifiers. But eyeglasses were more than merely difficult to explain; they were anathema to learned discourse. In addition to sharpening eyesight, spectacles appeared to alter the images being seen, convex lenses enlarging them, concave lenses shrinking them. Sometimes, given poorly made lenses, eyeglasses would distort shapes, or change the colors an unaided eye would detect. The conclusion was obvious: eyeglasses deceive the eye and degrade the central function of vision—to see the truth directly. Such trickery belonged to conjurers, not to those whose business it was to trace the evidence of the divine through science.

Which meant that spectacle makers had the field to themselves, along with a growing market. Their customers came to include even churchmen—the catalogue of an Italian bishop's possessions in the fourteenth century lists a pair of silver-framed eyeglasses—while writers found their spectacles incredibly valuable. Petrarch acquired his pair in the last decade of his life, sometime during the mid 1360s. By the sixteenth century they were common enough that Shakespeare committed one of the great anachronisms in literature by placing spectacles upon old men's noses in *Coriolanus*—a play whose action takes place 1,700 years before eyeglasses appeared. Any lingering suspicion that spectacles might obscure the evidence of one's eyes seems to have been fully laid to rest by 1600. One of El Greco's most chilling paintings shows the Grand Inquisitor, required by his office to distinguish between truth and the lies of the Great

Deceiver, calmly staring out at the world through a pair of black-rimmed glasses.

Finally, suddenly, after three hundred years of spectacle making, someone—several people—made the leap of technological imagination. On September 25, 1608, local officials in Zeeland, in the southwest of the newly formed Dutch Republic, wrote to the national government in The Hague that a local spectacle maker named Hans Lipperhey had invented a "device by which all things at a very great distance may be seen as if they were nearby." Lipperhey had at last done what seems so obvious in hindsight: he took two lenses from his shop, one convex and one concave, held them up together, noticed that they magnified objects at a distance, built a little tube to hold them and then realized that he had made something new: a telescope. Lipperhey applied for a patent for his invention but found himself stymied by the second great mystery of the telescope (the first being what took so long). Lipperhey's great discovery was not his alone—everyone in the Netherlands had seemed to hit upon the idea, apparently independently, all at the same time. Before the Dutch government could decide on Lipperhey's patent application, two other men, Jacob Metius and Zacharias Janssen, claimed the same invention—while at least one peddler was already offering a telescope for sale at the autumn market in Frankfurt, three hundred miles from The Hague. In the end, Lipperhey lost out: the Dutch government told him, in effect, that he had made a wonderful, useful thing—but that it was too easily imitated to be patentable.

At about the same time (the precise date is unknown) the same Janssen who helped frustrate Lipperhey's claims seems to have created (or claimed the credit for) another device, an instrument that would look down rather than up, in instead of out. The invention of the microscope seems to have been just as serendipitous as that of the telescope—and may have been a chance by-blow of the initial accident. Once the spectacle makers of the Netherlands realized that combinations of lenses could enlarge distant images, trial and error would have led one of them to a system that could magnify objects close at hand. Whether or not he was first, sometime during the early years of the seventeenth century Janssen combined two convex lenses in a tube to build a microscope that could magnify its subjects about nine times.

The earliest example of a scientific study made with a micro-

El Greco's portrait of Cardinal Fernando Nino de Guevara, head of the Spanish Inquisition.

scope came in 1625, when a Francesco Stelluti published his *Descrizzione dell' Ape* (*Description of a Bee*), complete with drawings made with the aid of a microscope. The continued development of the microscope proceeded quietly during the seventeenth century, with little public notice. Pictures of honeybees at low magnification were interesting, even amusing, but they didn't expose anything foreign to ordinary experience. Such images seemed no different from what could be seen if one simply squinted at an object from close range. Yet as microscopes improved over the course of the seventeenth century, they became the weapons in a kind of guerrilla campaign waged on behalf of the scientific revolution. Between 1608 and the 1680s a number of investigators pursued the sustained observation of the natural world through high-powered optical instruments. While, and partly because they did so, the object, the intent of scientific inquiry had shifted, marking the invention of what we can now recognize to be modern science.

But before the microscope could reveal its corrosive vision of the invisible within, seventeenth-century scientists had to remove the lingering distrust inherited from medieval natural philosophers: the idea that what was seen through a lens was somehow a lie. That battle was fought and won by the microscope's more famous sibling, the telescope. In 1608, Galileo Galilei was a forty-four-year-old professor of mathematics at the University of Padua, within the Venetian Republic. He was capable enough to have held that job for sixteen years, serving at the pleasure of the Venetian senate, and had completed much of his study of motion. By that time he would have reached the famous conclusion generated by the (almost certainly apocryphal) experiment at the Leaning Tower of Pisa, that bodies of different weight fall at the same rate. Galileo's father, Vincenzio Gallilei, was a musician—a lutenist and composer— of note, who was an early important exponent of using acoustical experiments to test the claims of music theory. Galileo, raised at this intersection of music and science, came by his experimentalist's passion early. But in 1608, he was still an ordinary professional academic, not terribly well known, and chronically broke. (He had both to supply his sisters with dowries, and to support his mistress and three illegitimate children.) He supplemented his rather meager academic salary by taking in boarders and running a successful sideline as an instrument maker, selling surveying tools, calculating machines, and other devices.

85

Intent as he was on juggling research, teaching, and the entrepreneurial life, Galileo failed to pick up on the first hints of the telescope's existence. He most likely heard talk of the new device in late 1608, when a friend of his in Venice received word from northern Europe, but rumors of wonderful inventions were common enough, and they usually proved to be false. Galileo himself recalled learning of the telescope in May 1609, when he was told that spyglasses that could magnify objects about three times were being sold in the market at Paris. The first night after he heard that report, Galileo later claimed, he returned from Venice to his workshop in Padua, acquired some spectacle lenses and built his first telescope.

That device was no better than the commercially available spyglasses then offered by peddlers throughout Europe. But Galileo applied his skill as instrument maker to a series of experiments to improve on the device. By August, he had succeeded in building a nine-power telescope, which he donated to the Venetian Republic, extolling its use in war: "This is a thing of inestimable benefit for all transactions and undertakings, maritime or terrestrial, allowing us at sea to discover at much greater distance than usual the hulls and sails of the enemy." Galileo noted that the telescope's many uses would be "clearly manifest to all judicious persons," and the judicious senators of Venice rewarded his donation and his discreet plea for improvement with a lifetime appointment at the university, along with a doubling of his salary.

At this moment, Galileo still appears as an ordinary, if talented, man of his times. He looked across, not up, extolling his telescope as a device with which to see better what was already there. But after the first rush to create a product to sell, Galileo began to develop a still better instrument for his own purposes. By November he had succeeded in building a telescope that could magnify twenty times, and in January he reported that he was nearing the completion of a thirty-power instrument.

The basic system he used, now called the Galilean telescope, was identical to that of Lipperhey, combining a weak convex lens with a strong concave lens. Except for opera glasses, this combination is not much used today, but it is an ideal design for a simple astronomical telescope. The Galilean telescope is designed to image objects at infinity—which in the language of optics means an object far enough away so that the rays of light it emits seem to be traveling in parallel lines as they reach a lens or the human eye. (For most contemporary camera lenses, for example, "infinity" begins about thirty feet away, the distance at which objects at infinity come into sharp focus.) The convex lens then bends the light, and would produce a sharp image where those rays of light converge, at the point called the focus of the lens, at a distance from the lens called its focal length. (The position of the focal point is determined by both the amount the lens curves and the refractive index—how much the material used to make the lens slows light down.)

In a Galilean telescope, though, the concave lens catches the image produced by the convex lens ahead of that focal point. Concave lenses are diverging lenses, which means they bend light outward. In the Galilean system, the concave element redirects the inward-bending light from the convex lens into a beam of perfectly parallel tracks, which produces an image of the object that the human eye perceives as being in focus at that same optical distance of infinity. The magnification such systems achieve is determined by the focal length of the convex lens divided by the focal length of the concave lens. To increase the magnification of his telescopes, Galileo used relatively mildly curved convex lenses with long focal lengths, in combination with sharply curved, short-focal-length concave lenses.

Given the quality of the lenses available to him, Galileo reached the maximum magnification he could hope to achieve with the twenty- to thirty-power instruments he produced after six months of effort. As he increased the strength of his telescopes Galileo also worked to improve the clarity of their images. All lenses suffer from a variety of optical effects that reduce the sharpness of the images they make. Rays of light shining directly on a spherically curved lens do not all meet at precisely the same focus point, causing a blurring called spherical aberration. A distortion called chromatic aberration occurs because light of different colors is refracted by different amounts in a lens, creating a rainbow halo or smear around the focused image. Both effects are worse at the edges of lenses, and Galileo found he could minimize the distortion by using just the central area of a larger lens, allowing less light through but bringing what light his telescope could gather into a far sharper image than his competitors could produce. In January 1610, Galileo fitted his telescopes with cardboard rings to block the light reaching the forward, convex lens in what was the earliest version of the system of aperture stops—familiar as f/stops in modern camera lenses. The higher the f/stop number—f/18, say—the smaller the hole in the ring, and the less light that gets through the lens. A lower number—f/1.8, a common maximum aperture for many camera lenses—corresponds to a larger hole and more light. Galileo's system worked out to about an f/50 lens—a very "slow" instrument, letting a small percentage of the available light through.

It was enough. Galileo began his career as an astronomer on the night of November 30, 1609. He was not the first to use a telescope to look beyond the earth. Thomas Harriot had already produced the first drawing of the moon made with the aid of a telescope during the summer of that year. But through his weak, six-power telescope, Harriot's moon looked much the same as one seen with the naked eye. By contrast, from the first night he looked upward, Galileo recognized that his far more sophisticated telescopes were carrying him into undiscovered country. He noticed first irregular, mottled patterns pockmarked on the lunar surface, and saw that the line between the dark and illuminated portions of the lunar disc twisted into a crooked, jagged track instead of following a clear, straight path. The moon revealed within the Galilean telescope could not be the smooth sphere required by cosmological systems that saw the heavens as perfect. Instead, Galileo recognized that the patchwork of light and shadow meant that the moon resembled the earth, covered with mountains and valleys. He even managed to arrive at a measurement of the height of the lunar peaks—about four miles or twenty-one thousand feet. (The highest peaks on the moon actually stand in excess of thirty thousand feet.)

In that first observing run, which ended on the night of December 18, Galileo, like other observers, saw many more stars with his instrument than he could with the unaided eye. He also sought to study the planets. He was able to see that, while the stars remained simply points of light in his telescope, the planets appeared as tiny discs. But more detailed observations proved difficult. In the weeks around the new year of 1610, Venus rose in the morning and was swiftly obscured by dawn; Mercury, rising and setting close to the horizon, was, as always, hard to spot; and Mars and Saturn lay far from earth on the opposite side of the sun. That left Jupiter.

As night fell on January 7, 1610, Galileo shifted his attention from sightseeing on the moon to begin the first detailed exploration of what the ancients had dubbed a wandering star. "At the first hour of the night," he wrote, "when I inspected the celestial

Galileo's drawing of the moon, based on his telescopic observations, from the first edition of his *Sidereus Nuncius*.

constellations through a spyglass, Jupiter presented himself. And since I had prepared for myself a superlative instrument, I saw (which earlier had not happened because of the weakness of the other instruments) that three little stars were positioned near him, small but yet very bright." Those stars fascinated Galileo, because, he wrote, they were brighter than other stars of the same size, and because they fell on a curiously straight line running through Jupiter—two to the east of the planet, with the third lying to its west. Consequently, Galileo returned to Jupiter the next night. The planet was traveling in retrograde motion in early January—that is, it moved from east to west against the background of the fixed stars—so Galileo expected to see Jupiter lying to the west of the "stars." Instead, he saw that the planet had slipped to the eastward of all three. He first thought that his astronomical tables were wrong and that Jupiter moved in proper, not retrograde motion. (All the planets move around the sun from west to east relative to the fixed stars, the direction of "proper motion." But the inner planets traverse their orbits more quickly than do the outer ones; as the earth overtakes the outer planets, those planets will appear to travel backward against the sky, following a retrograde path.) The test would be Jupiter's next position, but clouds gathered on January 9. On the tenth, though, the skies cleared. Once again, Galileo sighted Jupiter through his telescope and saw that "only two stars were near him, both to the east. The third, as I thought, was hidden behind Jupiter. As before, they were in the same straight line with Jupiter and exactly aligned along the zodiac. When I saw this . . . I found that the observed change was not in Jupiter, but in the said stars. And therefore I decided that henceforth they should be observed more accurately and diligently."

Galileo watched again on the eleventh, twelfth, and thirteenth. On the twelfth he saw a small star he had not viewed before, and on the thirteenth for the first time he saw four stars together, aligned on either side of Jupiter. Bad weather returned on January fourteenth, but, in the third hour of the night of the fifteenth, Galileo saw all four again, lined up to the west of Jupiter. Four hours later only three stars were visible, and in another hour, two of the three had moved noticably closer together. The explanation was inescapable: the "stars" dancing so swiftly around Jupiter were wanderers themselves—planets in Galileo's terms, moons in ours—circling Jupiter as our own moon orbits the earth.

Galileo's own account of his sudden moment of recognition is a model of scientific modesty. "I therefore arrived at the conclusion, entirely beyond doubt, that there are . . . stars wandering around Jupiter." Wanderers around Jupiter? Entirely beyond doubt? In this one sense, Galileo is a thoroughly untrustworthy witness. He cannot have been so calm on the night when he saw through his telescope a universe entirely unsought and undiscovered for more than two thousand years of staring at the sky. As a young man, Galileo himself had accepted the old, Greek Ptolomeic description of the cosmos: immutable, complete, the perfect bodies of heaven orbiting around the earth that rested at the center of all things. Galileo was a polymath: his studies of motion, of mechanics, his development of the experimental method, and above all his recognition that the laws of physics must be ex-

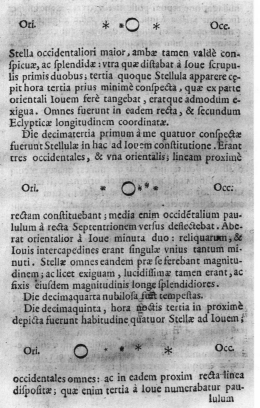

A page of *Sidereus Nuncius* showing Galileo's drawings and notes of his observations of Jupiter's moons.

pressed in mathematical language, all profoundly shaped the creation of modern science. But the epiphany came when he not only saw but accepted the consequences of his sight. It takes a leap of imagination to catch even an echo of the change that rang then, but as it chimed Galileo crossed the threshold between ancient and modern, and the world around him struggled to keep pace.

Galileo completed his observations by March 2, 1610. He published his results on the thirteenth, complete with drawings of the lunar surface and Jupiter's moons (which he named the Medicean stars, in an effort to recruit patronage from Cosimo Medici, grand duke of Florence). The first printing of 550 copies of the pamphlet *Sidereus Nuncius* (*The Starry Messenger*) sold out within a week. Almost immediately, the attacks began. The first to try was Galileo's bitterly jealous rival, the Bohemian astronomer Martin Horky. He was present when Galileo demonstrated his telescope for the faculty at the University of Bologna. Several faculty members had great difficulty picking out the Jovian moons, and Horky concluded that the alleged new planets simply did not exist. The fault, he concluded, lay not with the stars, but with Galileo. At his turn at the telescope, Horky said he saw four stars, just like the Jovian moons, around the third star in the tail of the Big Dipper. Clearly, Horky concluded, whatever the telescope's value for observations on earth, "in the heavens, it deceives." Of Galileo he wrote: "Full of himself, he hawked a fable."

Horky found himself almost alone in his categorical rejection; Johannes Kepler, Europe's most famous astronomer, accepted Galileo's claims immediately, despite the lack of a telescope of his own. But Horky's extravagant criticism pointed toward a deeper problem that Galileo had to overcome: the question of whether what the telescope saw was real. Spectacles could be quietly tolerated as philosophical eyes aged and vision dimmed. An instrument that brought back news of unknown worlds had to be confronted. Orthodox medieval Aristotelian doctrine was clear: the unaided sight of the human eye was the only true source of visions of the material world. To men who followed Aristotle, Galileo's moons, his gift to the Medicis, existed only inside the glass of his telescope lens.

Galileo's own skill as an observer worked against him. His telescopes were difficult to use. They were more than a meter (yard) long, with apertures stopped down to a couple of centimeters (an inch, or less), which meant they had to be pointed precisely at the object of interest—any shake or tremor would make observation impossible. Their field of view was small enough to make finding one's way around the sky a very delicate task, and some who tried to use the new devices reported quite honestly that they could not see the objects Galileo claimed were there. Throughout the spring of 1610 the moons of Jupiter had a will-o'-the-wisp character: only Galileo and those whom he personally helped with his telescope could convince themselves that they saw the Jovian satellites—and then, at the end of May, Jupiter disappeared behind the sun.

To a cautious thinker, such a state of affairs meant that the reality of telescopic images had to be open to question. Dancing planets only some could see would seem to be the definition of a fantasy. A lens could clearly alter images; did it invent them? To put the issue to rest observers other than Galileo would have to see what he had described. The chance came with Jupiter's reemergence from behind the sun at the end of July. Kepler saw the moons late in August, and other confirmations followed. Each new report increased the pressure on the scientific establishment of the church to decide whether to accept the evidence of a telescope as real. Finally, early in 1611, the head of the Vatican's Collegio Romano, Cardinal Robert Bellarmine, charged the scholars under his authority to settle the issue once and for all. On March 24, 1611, the four leading mathematicians in the employ of the Catholic church reached their conclusion: each of the discoveries so far reported by Galileo Galilei was valid. An instrument had produced new, true visions; the telescope extended human sight beyond the reach of the human eye.

Galileo, of course, paid dearly for being right. In part, Galileo would become an innocent victim of a power struggle that had little to do with him or his science. But nonetheless, the Catholic church asserted as a matter of doctrine that all the cosmos circled the center of creation, the earth—and as a young man Galileo had taught his students accordingly. But what he learned from his studies of motion, and especially what he saw with his telescope, convinced him of the truth of the theory developed by Nicolaus Copernicus more than fifty years before, that the earth moved, oribiting the sun, along with the other planets. Galileo first ran into trouble in 1615, when he was warned not to defend Copernican ideas. The crisis came later, in 1632, when he finally took up the question in print in his *Dialogue Concerning Two World Systems*. The defense of the wrong theory had earlier attracted the church's collective attention; the direct challenge to church authority cast a gauntlet before the Inquisition.

The church understood what Galileo did not. Galileo's use of evidence gained with his telescope implied access to truth, independent of revelation, awarding the individual observer the power to interpret the book of nature, which had heretofore been the exclusive right of the Catholic church. The famous trial followed, ending, inevitably, with Galileo's condemnation, as part of the inexorable calculus of power. He confessed and recanted publicly, swearing: "I abjure, curse, and detest the aforesaid errors and heresies, and also every other error, heresy and sect whatever contrary to the Holy Church."

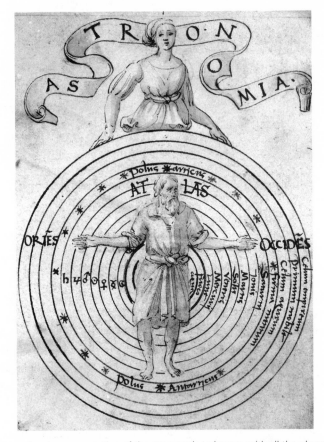

A late medieval representation of the geocentric universe, with all the planets and the sun orbiting the earth.

Galileo died in 1642, nine years after his trial, still under house arrest. His punishment effectively chilled interest in the Copernican system for a time in lands under Catholic control. But the telescope remained at large and people could look through it, if they kept their incautious conclusions to themselves. The urge to see, and see farther, remained unchecked in the doctrinal battles.

The story of the microscope in the seventeenth century is a quieter, more contained one than the high drama of Galileo's discoveries, but what it revealed completed what the telescope had begun. Together, the two instruments forced those who used them into a kind of looking and seeing radically different from what had gone before, one we now recognize as the central perception of modern science.

At first, though, the microscope was virtually ignored, and rightly so. The earliest versions were low powered and extraordinarily cumbersome. Galileo himself made microscopic observations and boasted of seeing "flies as big as lambs." But his instrument stood nearly as high as a man, had a tiny field of view, and was sufficiently difficult to bring to bear on any subject that Galileo only sporadically pursued its possibilities. Zacharias Janssen's ornate and decorative examples were more furniture than scientific apparatus. The Dutch ambassador to Louis the XIV's court described one of them: it was "almost a foot and half long, made of gilt brass two inches in diameter, mounted on three dolphins of brass, on a base of a disk of ebony."

Such early microscopes used a system of two lenses, both of them (unlike the Galilean telescope) convex pieces of glass. The problem the inventors of the microscope faced was how to bring into focus objects much closer than infinity. Point sources radiate their rays of light spherically, and shy of infinity light from such objects is not parallel but diverging, each ray bending outward. To bring the microscopic world into focus, the instrument builders had to bend that light back to form a sharp, magnified image that the eye could detect. Janssen and others solved it by using a so-called biconvex lens, one that curved outward on both sides, sharply enough to have a very short focal length. The object to be studied would sit exactly one focal length away from the lens, which would bend the beam of light from that object sharply enough to form an image of the object in a short enough distance behind the lens. The second lens could then intercept the beam of light behind that image plane, refracting it once again, magnifying it further. Such instruments faced the same technical problems that plagued the telescope—chromatic and spherical aberrations blurred images, and the sharply curved lenses needed to create images at short focal lengths were difficult to grind precisely. The quality of available lenses limited the maximum magnification early-seventeenth-century instrument makers could achieve. Janssen's microscopes could magnify as much as nine times, and by the mid-1600s the best commonly available microscopes had improved to only twenty-five- to thirty-power instruments. That was more than enough for Stelluti's bees (wonderfully monstrous views of insects were one of the most popular early microscopic subjects) but not to produce any image to compare with lunar mountains or moons around Jupiter.

Still, the accumulation of details of the known world could be an end in itself. In 1662 the infant Royal Society of London hired Robert Hooke, then just twenty-seven years old, to fill a newly created post: curator of experiments. The choice was a good one for both sides. Hooke wielded a fantastically inventive, intuitively precise mind, one that would make pioneering observations across the whole breadth of scientific inquiry, from physiology to physics. He was a skilled instrument maker who had already built his own microscopes, considerably refining the original designs. (He built, for example, a sophisticated system for illuminating the object under study, using a globe filled with clear brine that refracted light from a candle for a convex lens that could focus the beam onto a pin that held the object.) And perhaps most important, Hooke was lucky: in the Royal Society he found a home precisely suited to his talents, one that emerged just as he entered his scientific maturity.

The Royal Society, formally the Royal Society of London for the Promotion of Natural Knowledge, filled a need made obvious by Galileo's brush with the Inquisition a generation earlier. From the fall of Rome to the seventeenth century, Europeans who studied nature made their homes in institutions created for other purposes—those of the church, primarily, or of the cities, anxious to preserve their independence from feudal control, or within clubs dedicated to private amusement. The Royal Society, governed by scientists, free from any outside support or interference, gave its members an intellectual independence from any authority but their own. Hooke was a poor man—unlike many other prominent scientists, he had to work for a living. In any time prior to his day he would have had to seek a patron and suffer the constraints that patronage imposes. (In fact, before the Royal Society beckoned, Hooke himself worked under the patronage of Robert Boyle.) But in the Royal Society Hooke became one of the first true professional scientists, possessed of a freedom of inquiry unanticipated half a century earlier. He was paid simply to examine nature, to look wherever his curiosity took him.

The Royal Society began receiving the benefit of Hooke's observations almost immediately. Beginning on April 8, 1663, Hooke delivered a new microscopic study to the society's members at each of their weekly meetings. Hooke's microscope could magnify objects more than forty times, strong enough to produce an extremely detailed view of ordinary objects. First up was a picture of wall moss. He drew a detailed sketch of the plant's anatomy and used his instrument to test at least one hypothesis about the plant—that it could grow directly on such surfaces as stone or brick—determining that "I have found it growing on marble and flint, but always the microscope, if not the naked eye, would discover some little hold of dirt in which it was rooted." Next came a study of cork. Hooke found the honeycomblike structure that gives cork its light weight and yielding texture. He called the individual compartments of the comb "cells"—the first use of the word in biology—and correctly described the role of cellular construction in making cork light, compressible, and a good air-tight seal for wine bottles. Examinations of the edge of a razor followed, along with a look at silk, studies of stones and minerals, of the stings of nettles, or pores in petrified wood. He drew gloriously terrifying pictures of leeches preserved in vinegar, spiders, dragonflies, gnats, and moths—and perhaps most famously produced an enormous, lovingly detailed, whole-body portrait of a flea.

In all of these Hooke achieved a precision and a then-unequaled view of fine detail. In each case, in addition to producing first-class drawings of the magnified objects, Hooke pursued a detailed discussion of the meaning of every structure observed. But none of his microscopic excursions yielded a fundamentally new insight, and none was meant to. Galileo's telescope studies had produced a book, and Hooke followed suit—publishing *Micrographia* in 1665. Hooke set out to write a bestseller (he needed the money, if nothing else), and he succeeded. Set down in layman's English, elaborately and beauti-

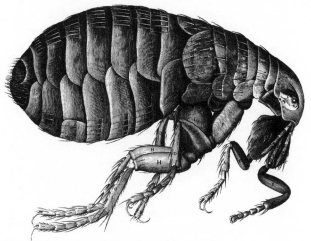

Robert Hooke's engraving of a flea, from *Micrographia*.

fully illustrated with thirty-eight engravings, *Micrographia* made an immediate sensation—the writer Samuel Pepys bought one of the first copies available, proclaiming that it was "a most excellent piece, and of which I am very proud." The illustrations were the prime selling point—editions of the engravings on their own were sold throughout the eighteenth century, and Hooke's meticulousness as a draftsman kept his observations in biology textbooks well into the nineteenth century.

But while Hooke looked for just this kind of commercial triumph, the carnival lure of giant insects and whole landscapes glimpsed on a sliver of stone that gained him his audience also served a larger purpose. Where *Sidereus Nuncius* recorded revolutionary discoveries, Hooke used his book to make a more subtle claim for a revolution in method. He wrote that "as for the actions of our senses, we cannot but observe them to be in many many particulars much outdone by those of other creatures . . . all the uncertainty and mistakes of human actions proceed either from the narrowness and wandering of our senses, from the slipperiness or delusion of our memory, from the confinement or rashness of our understanding." Given the deficiences of human capacity, in fact, Hooke argued that "the first thing to be undertaken in this weighty work is a watchfulness over the failings and an enlargement of the dominion of the senses." This Hooke proposed to do by "supplying [the senses'] infirmities with instruments, and, as it were, the adding of artificial organs to the natural." And such could now be done, he told his audience, because "by means of telescopes there is nothing so far distant but may be represented to our view; and by the help of microscopes, there is nothing so small as to escape our inquiry; hence there is a new visible world discovered to the understanding." He concluded: "The truth is, the science of nature has been already too long made only a work of the brain and the fancy: it is now high time that it should return to the plainness and soundness of observations on material and obvious things." The pursuit of such certain knowledge was the central function of the individual scholar and of the collective intelligence of scientists banded together. Already, Hooke claimed, he and his colleagues had found "some reason to suspect that those effects . . . confessed to be occult, are perform'd by the small machines of nature, which are not to be discerned without these helps."

These words mark the distance traveled in the fifty years since Galileo glimpsed Jupiter's moons. Hooke, and by extension the new scientific establishment gathered in the professional societies, could simply ignore the arguments that dogged Galileo—Hooke wasted no time, for example, with the by-then forgotten argument over whether lenses lie. Instead, he pursued bigger game, arguing that the use of instruments that altered human capacities provided the best hope of producing the discoveries that could explode the errors and fallacies that were the inevitable by-product of the old way of doing science, using only "the brain and the fancy." What could be seen was real, true; what could not was irrelevant. The imperative was to see more, as much as possible. Such observations would then serve the Royal Society's larger purpose: "The end of all these inquiries, they intend to be the pleasure of contemplative minds, but above all, the ease and dispatch of the labours of men's hands."

This has the sound of a manifesto. Hooke used *Micrographia* to trumpet the emergent power of a still-forming scientific method. What was sold as a report on the discoveries made possible with a microscope actually told the story of how such discoveries could be made. It was propaganda, in the end, and it was a recruiting poster too, one that reached the single volunteer it needed to catch. By the late 1660s, the mercantile wars between England and Holland were ebbing, and goods began to pass with some regularity across the North Sea. The best circumstantial evidence suggests that sometime around 1667 a copy of *Micrographia* reached the thriving market center of Delft, where it seems to have caught the eye of the draper Anton van Leeuwenhoek. By the time he was finished, Leeuwenhoek had done what Hooke could only propose, using his instruments to map Hooke's half-glimpsed "new visible worlds."

Leeuwenhoek's breakthrough came from his extraordinary abilities as an instrument builder. He took a design proposed by Hooke and produced microscopes—more than five hundred by the end of his life—that at their best were at least six times as powerful as any other instruments in use. Hooke himself had used the traditional microscope, similar in its essential principles to the original instrument invented in Holland around the turn of the seventeenth century. But in *Micrographia* he mentioned that he had tried several other types of microscope, and he described one such in great detail. Take a broken piece of glass, he wrote, and melt it over a fire until it forms a glass drop or bead. Use a needle to prick a hole in a plate of metal; mount the bead in the plate with sealing wax and the resulting device "will both magnifie and make some objects more distinct than any of the great microscopes." Nonetheless, Hooke rejected the device because "these, though exceedingly easily made, are yet troublesome to be used."

Hooke objected to the small size of the device in particular, and probably had little real interest in the delicate job of producing good lenses, drop by drop. His life was extraordinarily crowded in the early years of his career, his duties to the Royal Society competing with his own appetite for scientific investigation and his hunger for both fame and money. In any event, he purchased the lenses he used in his larger microscopes rather than grinding them himself. But Leeuwenhoek was a different kind of man, with an ambition wholly his own.

Leeuwenhoek was born in 1632 in Delft, son of a basket maker and his wife. He received some schooling but moved on to Amsterdam when he was sixteen to learn his trade, becoming a linen draper. Six years later he returned to Delft, married the daughter of a cloth merchant, and set up shop. He was apparently a successful and respected businessman, not the semiliterate, impoverished cloth seller of legend. He trained as a surveyor, knew some mathematics, and by 1676 was a sound enough merchant to be chosen to make sense of the tangled money problems left by the painter Jan Vermeer's early death. To outward appearance, Leeuwenhoek was the model of a rising Dutch burgher.

Then, in 1673, when he was forty-one years old, Leeuwenhoek sent a letter to the Royal Society describing his microscopic investigations of a louse, a bee's sting, and a form of mildew. He came to the Royal Society commended with somewhat faint praise—the Dutch diplomat Constantijn Huygens wrote a letter of introduction to the English scientists in which he called Leeuwenhoek "a person unlearned both in sciences and languages, but of his own nature exceedingly curious and industrious." Leeuwenhoek's letter set off an avalanche of papers. Between 1673 and his death in 1723 he sent nearly two hundred microscopic studies to be published in the Royal Society's journal, the last coming days before he died, describing "corpuscles in blood" and the structure of the dregs of wine.

Why Leeuwenhoek did it—how he came to be interested in microscopes, well established in his life as he was—remains one of the great mysteries of his career. His character tells part of the story. He was, by his own testimony, obsessive, secretive, relentless, and possessed of remarkable powers of concentration. He refused to reveal his techniques, either of instrument manufacture or of observation; he wanted, it seems, to see for himself, and to see first whatever lay hidden beneath the threshold of human vision. But nowhere in his letters does Leeuwenhoek reveal what set him off in the first place. The most plausible suggestion comes from the detective microscopist (his own phrase) Brian Ford, who wrote the book *Single Lens*, an excellent account of Leeuwenhoek's work. He points to two critical clues. Leeuwenhoek's classic microscope design uses a glass bead mounted in a small metal plate, with a screw-mounted specimen holder—almost exactly the microscope Hooke described then dismissed. And in several of his letters Leeuwenhoek responds directly to claims made in *Micrographia*. It is a guess, but seemingly the best guess, that Hooke's promise that new worlds existed for the taking struck the middle-aged Leeuwenhoek with such resonance that it awoke the passion he indulged for another fifty years.

The root of his success is easier to trace. The simple microscope that Hooke proposed and Leeuwenhoek used had one compelling advantage over the compound microscopes (two or more lens instruments) in general use. A simple microscope is essentially just a magnifying glass. The object to be examined sits at one focus of a lens that is convex on both sides. The magnification achieved depends on the focal length of the lens— the shorter it is, the better. The human eye has a focal length of about 10 inches or 250 millimeters—objects closer in blur away. A lens with a short focal length creates a focused image of objects that would normally be invisible. The lens, as it were, pushes back the object, enlarging it as it goes. The actual magnification is determined by the ratio the focal length of the eye divided by the focal length of the lens. Thus, a 25-millimeter focal length lens will yield 10 times magnification, for example, while a 2.5-millimeter lens would magnify an object 100 times. It is easier to make a lens with a relatively long focal length, so the idea of combining several such lenses to maximize magnification seemed the most promising. But the usual spherical and chromatic aberrations would degrade the image at every step through the instrument, and any flaw in the lenses would multiply through the system. In practice, microscopes like Hooke's rapidly reached a limit of about 50 times magnification, beyond which instrument builders of the day lacked the precision to go.

Leeuwenhoek's great advance came as he realized that, with

enough care, he could take the simple magnifier—of the sort, as it happened, he would have commonly used to look for defects in cloth—and produce a single-lens microscope with the lens shaped so accurately that it would be more powerful than any multiple-lens system then available. The trick lay in a willingness to devote obsessive attention to the manufacture of the lenses. Leeuwenhoek reported that it took him ten years to perfect his methods—time no one but he, in effect, was willing to give to the cause. The best surviving Leeuwenhoek microscope contains a lens that is a single bead of glass so carefully shaped it can magnify objects 266 times.

With such an instrument Leeuwenhoek gained passage to a realm of phenomena as startlingly unexpected as the existence of moons around distant planets. The "animalcules" he saw in Berkelse Mere were the first microbes ever recognized by the human eye—and Leeuwenhoek saw and described them clearly enough to allow modern biologists to identify the species found in his sample. He would go on to observe dozens of species of microbial life, tracking down the presence of bacteria, for example, in samples of pepper and cloves. He studied the spermatozoa of different organisms and performed a series of experiments with rabbits and dogs, to track the movement of sperm up the reproductive

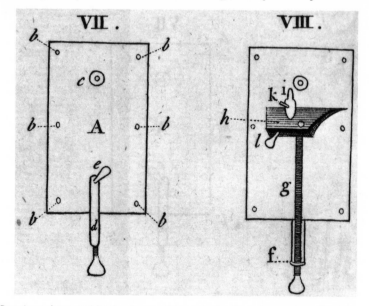

Drawings, front and back, of one of Leeuwenhoek's microscopes. The screw-mounted pin holds the specimen in front of the lens.

tract. He investigated the pesticidal properties of substances like sulfur and nutmeg. He analyzed materials, comparing, in one study, oak grown in France and Holland. (He was able to confirm that French oak was stronger than Dutch, a consequence, Leeuwenhoek argued, of the warmer weather and greater light that reached the southern trees.)

In all of his observations Leeuwenhoek displayed an extraordinary discipline. He interpreted all that he saw—in his description of the pesticide action of nutmeg, for example, he noted that the nut broken open killed mites faster than did any portion of the nut with an unbroken husk and concluded "that the vapor of the nut evaporates much more feebly through the husk than from the newly broken internal portion." But throughout his work he compelled himself to test his ideas against the evidence he could detect within the view of his microscopes. His most elegant work lay in his investigation of the idea of spontaneous generation, the doctrine that living organisms could emerge directly from inert matter. Hooke was a believer, but Leeuwenhoek, in a series of experiments with maggots, parasites, moths, and other organisms, demonstrated that for all the animals he could observe, the adults of one generation reproduced among themselves to create the next generation of young. Then he chastized those who sought what he saw as a supernatural explanation for a process that lay open to anyone with the wit to look: "Can any man in his sober senses imagine that the moth of which I have given this description, which is duly provided by nature with the means to propagate its species . . . can this moth, I say, adorned with so many beautiful features, be produced from decay?"

The essential character of Leeuwenhoek's work throughout, in fact, is the approach that lies at the core of the modern scientific sensibility. His was an endless search for mechanism: what happens, how it takes place, when it occurs, where it occurs. Ford includes in his account the story of Leeuwenhoek's treatment of the legend that black scraps of charred-looking material that turned up on sea shores was actually "heavenly paper," burnt as it fell from the sky to earth. Leeuwenhoek received a sample of this paper from the shores of the Baltic and from the start rejected any thoughts of the supernatural. Viewed through his microscope, the material showed the traces of its origins as an aquatic plant, algae (later work found the remains of waterborne microbes amongst the "paper" fragments). Leeuwenhoek reasoned that the paperlike plant leavings could be formed after flood waters receded, and algal mats lay exposed to dry in the sun. With that as his working theory, he performed the experiment that finished off the extraterrestrial hypothesis. He gathered some fresh algae, set it to dry in front of his fire, and produced some heavenly paper of his own.

End of mystery. Leeuwenhoek saw the traces of natural origins in the odd, blackened flakes, and then established, elegantly and efficiently, what events could have created them. In doing so, Leeuwenhoek evoked, almost invented, a way of seeing, a kind of vision, that has become the dominant form of the scientific gaze. The medieval eye, Roger Bacon's eye, was passive. Bacon looked at what passed before his eyes and stopped when he had seen enough to recognize the hand of God in nature. Leeuwenhoek, exploring the unimagined world his microscope revealed, entered that world, engaged it actively, intervened with his eye to expose its secrets to view, performing experiments to track down its governing mechanisms, dissecting it—literally—to reveal detail unimaginable with the evidence of the naked eye alone. The tradition of experimental science is so deeply embedded now that we accept unquestioningly the idea of breaking into a natural process to better understand its parts. But it requires a leap of vision to imagine that one could mimic the action of sea and sun with a scrap of algae and a hearth fire. Leeuwenhoek's extraordinary accomplishment derives not just from his ability to see ever-finer detail but from his acute skill at picking out the critical phenomena that could be exposed to deeper study, closer examination, or experimental test.

Such clarity of vision forms a striking contrast to the tangled mixture of acute observation and wild fantasy that marked the work even of such original thinkers as Bacon. And yet, the central thought that undergirded Leeuwenhoek's approach evolved directly out of beliefs that Bacon would have held. The old idea, that nature does nothing in vain, still had meaning for Leeuwenhoek—but with one critical difference from the medieval belief. To him it meant that nothing visible in nature is meaningless; everything that can be seen can be understood in terms of the natural processes that could create the phenomena in view through his lens. Bacon looked to nature to provide a truthful record of God's purposes; he wanted to understand why the world was made as it was. Leeuwenhoek sought to understand how his paper was made; how moths produced more moths;

which trees were stronger than others; how pepper created its spicy sensation; how microbes move; and so on and on through all his communiqués to the Royal Society. But in all those letters, he never asked the older question, "Why?"

The Franciscan Bacon existed entirely bound within the context of the church, pursuing his studies for his best notion of what its ends were. Galileo learned the cost of challenging the church's prerogative. But Leeuwenhoek was a townsman, a merchant, building his instruments and peering at his samples because it amused him, because he chose to. He lived in a society that had shrugged off the authority of a singular, centralized sytem of belief, and had access to a community that had not existed a century before, men whose common bond lay simply in their pursuit of material knowledge. Roger Bacon had despatched his best and most treasured work to the pope in Rome, who paid it little heed. The draper Leeuwenhoek was elected to the Royal

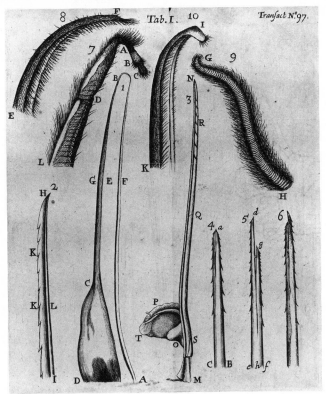

Leeuwenhoek's observation of bee stings and antennae, published in the *Philosophical Transactions* of the Royal Society.

Society of Hooke and Isaac Newton, in recognition of the value of his scientific work—and published his letters in its *Philosophical Transactions* for the general edification of all interested parties.

Critically, Leeuwenhoek, that ambitious man, clearly defined his ambition, limiting it in a way medieval scientists had not. Telescopes and microscopes, those "artificial organs," do not simply extend human sight. They narrow it, confining the field of view. Leeuwenhoek, squinting at the microbes swimming in the water of Berkelse Mere, could see a city in a single drop, but not the pond itself, not unless he pulled away from his lens. As his instruments improved, he could see more and more of less and less. Similarly, as modern science created itself out of its medieval roots, it narrowed its range, a consequence of banishing "why" from its vocabulary. Scientists of the new type looked for the laws that established the relationships between one observable phenomenon and another—correlations, not causes. Newton's law of gravitation, the most famous example, relates the force of gravitational attraction between two bodies to the mass of each and the distance between them. With the formula in hand, one can calculate the change in that force for any shift in mass or distance. One can predict the path of the space shuttle or explain—as Newton did—the elliptical shape of the orbits of the planets around the sun that Johannes Kepler had discovered. But nothing in Newton's law reveals why gravity works, who pushes or pulls, for what reason. This kind of science is descriptive—it establishes what happens, to the highest degree of accuracy possible—and its jurisdiction ends there. To achieve the rigor and predictive power of accurate description compelled the new science to pay a penalty steep enough to trouble many to this day: Galileo first, then the generation that included Leeuwenhoek and his correspondents at the Royal Society, abandoned the search for the knowledge that could lead man through nature to God. For these men, the kingdom of heaven still existed—but beyond their reach, outside of the focus of their augmented eyes and minds.

Newton himself understood perhaps more clearly than any of his contemporaries the meaning of this new approach to nature. Elected president of the Royal Society in 1703, he published his book *Opticks* the following year. What Galileo and Leeuwenhoek and many others developed ad hoc, by trial and error, Newton analyzed and systematized—and in the last, more speculative section of the work, he described the method all scientific investigations would henceforth have to employ:

> In Natural Philosophy, the Investigation of difficult things by the Method of Analysis ought ever to precede the Method of Composition. This Analysis consists in making Experiments and Observations, and in drawing general Conclusions from them by Induction, and admitting of no objections against the Conclusion but such as are taken from Experiments, or other certain Truths. . . . And although the arguing from Experiments and Observations by Induction be no Demonstration of general Conclusions; yet it is the best way of arguing which the Nature of Things admits of. . . . By this way of Analysis we may proceed from Compounds to Ingredients, and from Motions to the Forces producing them; and in general from Effects to their Causes, and from particular Causes to more general ones, till the Argument end in the most general.

The truths science could create must emerge inductively, that is, from the facts science could establish, and from nowhere else.

Despite such words, Newton himself possessed a vision of a divinely written book of nature. In the same passage, he added: "If natural Philosophy in all its Parts, by pursuing this Method, shall at length be perfected, the Bounds of Moral Philosophy will be also enlarged. For so far as we can know by natural Philosophy what is the first Cause, what Power he has over us, and what Benefits we receive from him, so far our Duty towards him, as well as that towards one another, will apppear to us by the Light of Nature." But the crucial shift had taken place, and Newton knew it. His God could be found in nature and its laws—not, like Bacon's, through them. In that God's presence, the scientists around Newton continued to construct instruments like the telescope and the microscope that with each improvement created new realms of knowledge. Medieval men could stop when they had achieved their object, when they had seen enough. The new, modern kind of scientist had no such luck; the inductive method required them to continue seek new evidence that could confirm or disprove their ideas, on and on, with no end in sight.

Anton van Leeuwenhoek died on August 26, 1723. He kept working to the very end—two days before his death he searched for gold in samples of sand sent him by the East India Company. His life spanned a time his own labors had helped to transform. One hundred years after Leeuwenhoek's death, Ralph Waldo Emerson would sound a triumphant note, a hopeful one: "So shall we come to look at the world with new eyes." Leeuwenhoek had long since built himself new eyes and found new worlds with them. He saw less and more than others had before him. So do we still.

INTERLUDE:

SOUND

AND LIGHT

Looking backward, with the scientific revolution fought and won, the difference between the old and the new seems total. The story can be told in the heroic vein, a tale of the seventeenth-century revolutionaries who fought those of traditional mind—their predecessors, their unmoved contemporaries—to decide what science ought to do and how it ought to do it. The triumphant general, in this telling, was Sir Isaac Newton, whose campaign culminated in the most famous victory in the revolutionary cause. His laws of motion and the universal law of gravitation demonstrated the power of descriptions of nature expressed in the language of mathematical relationships. Critically, the mathematization of the natural world allowed (and allows) the scientist to test claims of natural law against the evidence of experience itself—Newton's Method of Analysis.

This triumph of method completed the scientific revolution, narrowly conceived. But Newton himself made clear that his battle turned on another, older conflict over the status of experience, of experiment itself. The invention of machines that altered and extended the reach of human perception fostered a new realm of experience, an age of discovery of a piece with the extraordinary press of maritime exploration European sailors undertook over the same decades and centuries. What the scientists found served the cause of revolution, of course. Galileo's moons helped settle the debate over Copernican ideas once and for all, and Leeuwenhoek's relentless attack on the notion of the spontaneous generation of life steadily eroded the power of supernatural explanations for natural phenomena. But more important than the specific findings, the instruments that found new worlds offered Newton and his heirs the one thing they needed to carry their scientific method forward: the promise of discovery, the wherewithal to gather ever more experience, ever more precise data with which to check and test, modify, reject, and remake their mathematical abstractions.

Armed with the method and the tools needed to apply it, this new science with its narrowed intention to describe nature, instead of looking for its motives, moved to dispose of the superstition, appeals to ancient authority, the wild excesses of argument that had characterized the mixture of magic, geometry, legend, observation, and all the rest of the explication of nature within the medieval world. And yet there is no such thing as a complete victory.

The scientific revolution, that is, was created not simply by the acts of heroes but by a process that grew directly out of the centuries-long attempt to understand how to study nature. The musical cosmology of Pythagoras and his successors in classical antiquity created the cornerstone of the earliest scientific inquiry, the recognition that the seemingly random details of sense experience could conform to a broad, abstract pattern or law. The central drama of the High Middle Ages turned on the realization that new phenomena could be discovered—new music could be made beyond the confines of the old system of musical science— novel experiences that could be placed within the overarching order that governed nature. The critical shift that fostered the scientific revolution emerged here, with the realization that whole new categories of phenomena could be found with mechanical extensions of human senses. Such discoveries compelled the new scientists to abandon the quest for the order behind nature—God, for most of them—and seek instead an account of the patterns and relationships that exist within nature. The break with the classical and medieval past is there; but it is not a clean one, nor did the men who made the revolution see it that way.

As their eyes bent to the machines that could aid a newly focused gaze, some of these makers of modern science retained the ears that could detect an echo of the harmony of the spheres. Johannes Kepler's discovery that the planets orbit the sun along elliptical paths represented one of the most radical leaps of thought of the scientific revolution—breaking as it did with what was probably the most deeply held belief of ancient science, that celestial objects move only in perfect circles. And yet for Kepler the sky sounded with heavenly music; his head rang with it, and his mind rejoiced in the song it made.

Johannes Kepler, born in 1571, was one of the earliest and most determined converts to the Copernican idea that the planets revolved around the sun instead of the earth. But the Copernican cosmos begged the question of what arranged the planets into their particular order: What determined their distances from the sun and the speed with which they completed the transit through the sky? Kepler's first thought was grandly ancient: the planets' orbits were each perfect circles inscribed on the five regular polyhedrons of Greek geometry—the cube, the tetrahedron, the dodecahedron, the icosahedron and the octahedron (moving inward from Saturn to Mercury). Miraculously (it seemed to Kepler), the idea worked, generating the orbits of the planets with errors of less than 5 percent. Increasingly precise observation held out the hope of confirming the theory, and Kepler set out to take hold of the best collection of planetary data in the world, the hoarded knowledge of the Danish astronomer Tycho Brahe.

Tycho was the last of the great naked-eye astronomers, a rich man who had devoted his life to precision measurement of planetary motion. He both needed and distrusted the young Kepler with his sharper intellect, and assigned him the problem of accounting for the observed orbit of Mars—one that seemed irreconcilable with both the Greek and the Copernican schemes. The first discovery Kepler made was that his geometry of the heavens simply did not work. He tried more than fifty different combinations of circles to account for the planet's motion, but none fit the positions in which Tycho had found Mars. It took a decade, but Kepler finally uncovered the correct answer: no set of circles could work; Mars drew an ellipse across the sky.

Kepler himself knew the gravity of his claim. He wrote that he persisted so long in trying to find a circular orbit because the necessity of such perfect planetary motion "was taught on the authority of all philosophers, and is consistent in itself with metaphysics." He overcame the siren call of circles (as Galileo never did, in fact) only because of his essential commitment to the authority of empirical data over and above that of the philosophers; no theory could withstand contradiction by the facts. But though Kepler's laws of planetary motion accounted for the behavior of the solar system as observed, Kepler himself retained his original passion, to go beyond the description of what was seen to uncover the deepest patterns that nature could form. After publishing his determination of the elliptical paths of the planets, Kepler continued to work for a decade, reaching the triumph he had sought from the beginning with a discovery that left him, he wrote, "free to give myself up to the sacred madness ... free to taunt mortals with the frank confession that I am stealing the golden vessels of the Egyptians, in order to build of them a temple for my God."

What Kepler found as he penetrated more and more deeply into the subtleties of planetary motion was a new music of the spheres, as described in his book *Harmonice Mundi* (*The Harmonies of the World*). Kepler compared the speed of each planet at the point nearest the sun, when planets move their swiftest, with their velocity farthest away from the sun, when planets travel at their slowest. With the ratio of those two speeds, he constructed musical intervals. Mars, for example, covered a perfect fifth, from C to G, the ratio 3:2, while Saturn sounded out a perfect third. All the planets could produce glorious glissandos—slides up or down—as they intoned their way around their orbits, until each planet produced its own song. Together the solar system generated the glorious interwoven sound that confirmed for Kepler the truth of his system: "The movements of the heavens are nothing except a certain ever-lasting polyphony ... hence it is no longer a surprise that man, the ape of his Creator, should finally have discovered the art of singing polyphonically, which was unknown to the ancients, namely in order ... that he might to some extent taste the satisfaction of God the Workman with His own works, in that very sweet sense of delight elicited from this music which imitates God."

Kepler was ecstatic, with good cause. His musical astronomy does in fact work, and the planets do produce a set of tones that fit into the octave astoundingly well. Today we can hear what Kepler could only imagine in a recording made by Yale University's Willie Ruff and John Rodgers, who used electronic instruments to play the planets' songs, appropriately sped up to bring them into the range audible to the human ear. Nor was Kepler alone in sensing the harmonic perfection and pattern evident throughout nature. Newton himself pursued the clues of music throughout his career. His early work on the arithmetic of Pythagorean harmony led him to the fifty-three-note octave mentioned above, but he returned to a simpler scheme to account for the progression of colors revealed in the spectrum of visible light. He wrote in his earliest important work on the subject that "as the harmony and discord of sounds proceed from the proportion of the aereal vibration, so may the harmony of some colors and the discord of others ... proceed from the proportions of the aethereal [vibrations]. And possibly colour may be distinguished into its principal degrees, Red, Orange, Green, Blew, Indigo and deep Violet, on the same ground that sound within an eighth [an octave] is graduated into tones."

Newton pointed out that the seven colors—the six above and yellow—could be lined up with the seven notes—A, B, C, D, E, F, and G—of the octave. (He ignored the sharps and flats.) Red sounded the low note, while deep purple rang out the top of the scale. It was a serious business: Newton even faced the same problems of tempering his scale that organ builders and piano makers confronted. He tried to calculate the tuning of his light-octave by measuring the distances between each of the main colors, coming up with a couple of different possible scales, each of which fit the data closely, though not perfectly. Still, Newton felt that harmonic theory worked well enough to indicate that the true connection between sound and light was simply waiting to be discovered and was already evident in the empirical experience of color. As he noted in *Opticks:* "For some Colours, if they be view'd together, are agreeable to one another as those of Gold and Indigo, and others disagree."

Newton here contradicts his famous claim—"I feign no hypotheses"—committed by his pursuit of harmony to an assumption that shaped and directed his perceptions. Newton knew that there are seven notes to an octave. He looked at the spectrum and saw the seven colors still recognized as the seven shades of a rainbow. And yet the spectrum's hues blend smoothly from one to the next, a continuum, with bands of greater or lesser intensity and width, and it is only long use, custom, and the distant call of an ancient tune that persuades us that we see just seven colors and not the intermediate tones. Newton's music cast the spectrum he made with his prism into the pattern he already knew.

That faith in a preexisting pattern, one that can be found out in nature, dominated Newton's science. Despite Newton's clear explication of his Method of Analysis, with its requirement that experience undergird all scientific generalizations, he retained this one assumption which the method could not test, for it rested upon the belief that all the data gathered by the scientific method can be ordered, that it conforms to harmonies that are accessible to human reason. Newton's universe was arranged (he argued) by a divine hand into a form revealed by divine, natural laws. Once those laws were understood mankind could gain its closest knowledge of God. Newton himself seemed to believe that eventually all such rules could be known, as thoroughly as his law of gravitation: "Whatever reasoning holds for greater motions . . . should hold for lesser ones as well."

It had been the same for Kepler. His heavenly music provided the link between his mind and God's; his science was driven by the belief that such links existed throughout nature, that the patterns he expressed as harmony, music, did, in fact, exist out there in the real world. He wrote: "Those laws are within the grasp of the human mind; God wanted us to recognize them by creating us after his own image so that we could share in his own thought. For what is there in the human mind besides figures and magnitudes? It is only these which we can apprehend in the right way, and if piety allows us to say so, our understanding is in this respect of the same kind as the divine, at least as far as we are able to grasp something of it in our mortal life. Only fools fear that we make man godlike in doing so; for God's counsels are impenetrable, but not his material creation."

Such boasts echo those of his predecessors. But Kepler hit upon the one change that modern science had to make to avoid the trap of teleology, seeking to understand why God acted one way or another. God's counsels are impenetrable, he reminded his correspondent. Science could not uncover such reasons, which left only the material facts at hand. The presence of a divine author behind such material facts should simply assure the scientist that the God who created both human reason and nature would have made the latter in such a way that the former could comprehend it.

Hence Kepler's ecstasy at the sound, heard in his mind's ear, of celestial music: it was for him a revelation. His belief in the existence of such patterns created what remains an aesthetic of science; when he recognized a given order, it appeared to him beautiful. The beauty and elegance of his inventions served to reinforce his commitment to the idea that nature forms such patterns. It restored to science what Newton's apparent rigor seemed to eliminate, the reward of intuition, of insight, in a sudden glimpse of harmony. And for his part, following Kepler, Newton emphasized this same aesthetic as the guide to those who would understand the world. "It is the perfection of all God's works that they are done with the greatest simplicity," he wrote, adding: "They that would understand the frame of the world must endeavor to reduce their knowledge to all possible simplicity." This is the idea that made modern science possible: that any natural phenomenon can be understood within a framework of abstract, universal, *simple* laws.

It was the Greeks, of course, who first glimpsed the existence of abstract patterns repeated across nature, from the sound of music to the motion of the heavens. Medieval men used such patterns to identify the trace of the divine hand at work, seeking God through nature. Scientific revolutionaries sought the rules that gave rise to such patterns, whose every discovery renewed their faith in a rational God acting, literally, behind the scenes. Contemporary scientists, heirs to each succeeding generation, retain the central assumption that all their predecessors held in common. Nature must make sense—in modern language, to paraphrase the physicist Eugene Wigner, nature knows mathematics. That belief allows the scientist to recognize, to state the ever more elegant descriptions that capture the endless accumulating details of discovery. And as did the ancients, modern scientists, contemporary ones, build on this article of faith an aesthetic of science. We still believe that as we understand nature's patterns, we will recognize their truth in the poetry of their form, in the beauty contained within them. Einstein made the search for order an act of faith when he said: "God does not play dice with the universe." Werner Heisenberg, one of the inventors of quantum mechanics, in conversation with Einstein understood science in the language of art: "If nature leads us to mathematical forms of great simplicity and beauty—by forms I am referring to coherent systems of hypotheses, axioms, etc.—to forms that no one has previously encountered, we cannot help thinking that they are 'true,' that they reveal a genuine feature of nature." Stephen Hawking, writing with the modern physicist's bravado, even accounted for his quest for the single unified theory of the universe with the hope it would reintroduce the question Why? to science, writing that if it did so, "it would be the ultimate triumph of human reason—for then we would know the mind of God."

The enterprise of modern science, that is, comes to rest upon its own claim of faith, that the harmony we respond to corresponds with patterns that genuinely exist, which we encounter with the emotion that rewards the sight of beauty. Such faith lies outside the realm of proof. We assert, essentially, as Kepler did, that "our understanding is in this respect of the same kind as the divine." Divine or not, the assumption works, of course, or it has so far, which is what distinguishes this particular statement of the scientific credo from the older forms. Science since Newton has progressed enormously in the attempt to derive natural laws of increasingly great abstraction and the broadest possible generality. The grand pattern nature is presumed to make remains unknown, of course, despite Hawking's hint that we might be getting close—but the perception that it might exist remains as essential to the pursuit of scientific discovery as it ever was; it still goads us on.

And yet, despite the success of the enterprise in the last three centuries, the idea that the scientific conception of nature itself is beautiful does not seem obvious to many. Here eighteenth-century ears and eyes may have had the advantage over ours, for their perception of celestial harmonies resonated with the earthly harmonies of the music others made around them. Sir Isaac Newton and Johann Sebastian Bach can be seen as men on parallel tracks, men whose lives can illustrate each other's. Of the two, Bach was the happier, almost surely. Certainly he realized his vision at least as fully as did the English scientist, achieving again and again his astounding marriage of precision and passion. Newton may have listened to the harmony of the spheres, but we perhaps come closest to hearing its strains in a piece Bach wrote around 1715, the Fantasia and Fugue in G minor, styled "The Great."

The fugue is a technique rather than a fixed form. In the G-minor piece, one theme (the subject), its response (the countersubject), the theme again, another response, are each taken up by a separate voice of the organ. The lines speak to one another, variation succeeding variation to build an increasingly complex weave of all the elements. The melodies move in and out, until the pattern closes at the end, coming to rest on the original theme. As it unfolds, the piece creates an astoundingly vivid sense of an inevitable logic, combined with an exalting, soaring quality that evokes an older image, the sudden height of a gothic cathedral. It is an artistic effect born of the same aesthetic that animates the search for the ideal form in nature; Bach's "Great" fugue matches any of Newton's mathematical arguments in its logic and formal elegance.

Newton might not have welcomed such a comparison—for all of his musical curiosity, he was jealous indeed of his reputation— but Bach certainly understood the link between his work and the larger intellectual currents around him. Throughout his career he pursued the connections between music and the techniques of thought employed by the makers of the scientific revolution. In 1747, three years before he died, Bach joined a society of composers that included Telemann and Handel called the Corresponding Society for Musical Sciences, dedicated to rationalism in art and thought. Before joining, each initiate had to submit a composition that demonstrated his commitment to the expression of reason in music. Bach's choice was a set of canonic pieces for organ that constructed a pattern of melody through six parts, a cycle of sound orbiting a musical center in movements that grew ever more intricate, ever more complex: the music of the spheres compressed into the compass of an organ's pipes.

Now, two and a half centuries after Bach's death, celestial music has become a metaphor, a poet's way of expressing the ancient, lasting credo upon which our belief in the possibility of scientific discovery depends. Bach's works create a bridge between the language of poets and the language of modern science. The glory of his great organ pieces is that they feel "true"—they are mathematically precise, complete, coherent—while remaining pleasing, to say the least, to the ear. They provide the experience, that is, that Bach clearly intended for his audience: they allow us to hear what Heisenberg meant when he spoke of the bond between beauty and truth in science.

From the beginnings in Greece to the present day, such sounds as an organ may make, the sights to be seen with the naked eye or through a lens, have driven the attempt to comprehend our surroundings. The scientific revolution of the seventeenth century altered the sense of what could be understood, and what must remain a mystery. The search for divine wisdom ultimately gave way to the artist's quest for beauty, the musician's for harmony. Retained across the divide of revolution was the dream of reason: that the combination of the new method and the old sense of order and pattern would reveal, inevitably, the inner workings of nature. Just measure precisely enough, build theories elegant enough, and the reward would be the ultimate, true understanding of the universe. To listen now to Bach's great fugue in G minor is to hear a hymn of praise and a prayer: praise that the world possesses such glory as the sound of that melody; a prayer that reason's dream might be granted swiftly. It was not.

An Unbounded Prospect

IT IS A

GREAT

SECRET

Two dogs tumble in front of the gates of Trinity College, Cambridge. The engraving, from late in the seventeenth century, also shows a laborer carrying a bundle of sticks up the road, and two men, apparently dons, oblivious to dogs and the common man alike, locked in learned conversation. Over the wall can be seen one wing of the college, leading to the chapel, against which stands a small shed. That shed next of the house of God is most probably the place where one resident of the college would retreat to perform what he regarded as his most secret and holy work.

Isaac Newton was ever an intensely private man—it took the urging of Edmond Halley and the threat of a rival's work to persuade him to publish his mathematical account of nature, the *Principia*—but never was he more jealous of his secrecy than in his shed on the grounds of Trinity College, his laboratory, the site of his detailed study of the art of alchemy. For Newton, alchemy was "a more subtile, secret & noble way of working"— subtle, as it allowed him to peer into the processes of nature that govern growth, change, and life; noble, as it placed the alchemical adept on the path of imitating the creative power of God; and secret, necessarily, for such power would be too dangerous to share recklessly with the vulgar.

Thus, Newton pursued his alchemical research in almost complete solitude. He built his own furnaces, laying the brick himself. He acquired the tools of his trade—"bodyes, receivers, heads, crucibles &c"—and taught himself the basic operations of chemical investigations. Humphrey Newton (no relation), Newton's assistant for five years, recalled after his master's death, that for "about 6 weeks at spring and 6 at ye fall, ye fire in the Elaboratory scarcely went out." As he worked, Newton would "sometimes, though very seldom, look into an old mouldy book which lay in his elaboratory . . . titled *Agricola de Metallica* [*The Farming of Metal*—the last great Renaissance metallurgical text], the transmuting of metals being his chief design."

Newton kept his larger plans from his assistant, and warned those few he did talk with to keep their mouths shut. He wrote to the pioneer chemist Robert Boyle, for example, that alchemical knowledge was "not to be communicated without immense damage to the world." In his passion for obfuscation Newton followed an ancient tradition. Alchemists over the centuries obsessively chronicled their experiments, and just as obsessively recorded their findings, in codes, symbols, mystical formulas, and allegorical prose. Newton labored over that vast body of alchemical literature, seeking to tease out of its deliberately obscure texts the keys that would open to him the vault of arcane knowledge, including that oldest of hopes, the conversion of base metals—lead, iron, and the like—into gold.

Newton owned 138 books on alchemy, 8 percent of his collection, more than he possessed on any other even remotely scientific subject. His notebooks contain lists of alchemical symbols—seven different signs for Mercury, for example, and five for sal

Detail from an engraving of Trinity College showing the shack up against the chapel where Newton probably performed his alchemical experiments.

ammoniac, including one in the shape of a flying insect—all arranged to ease the comparison of one text with another. At the same time he wrestled with individual works word by word, looking for the clues he needed to complete his own experiments. In one of his notebooks, he copied a diagram of the philosopher's stone—the active agent alchemists used to cause the transmutation of base metals into gold—and in another note, he laid out his understanding of "an old priest's" experiment: "Put this mercury in a glass retort between two capps soe as it neither touches ye sides nor bottom of the capps, and with a good fire under and hot embers on ye top to keep ye heat of ye fires ye better for 40 hours, ye mercury will distill into a slimy mater hanging together. It will melt nothing but metall. This is the true aqua vitae, ye spirit so much desired of philosophy."

In this recipe there is the hint of a scientific experiment. Newton offers a clear and unambiguous procedure for manipulating mercury to produce a desired compound, one presumably open to anyone with a laboratory and sufficient curiosity. Newton's critical piece of apparatus was the still, a double-chambered vessel he used for the purification of mercury. The still is one of the oldest of alchemical/chemical instruments, dating back at least to the age when the Romans ruled Egypt. It served one of the basic aims of the alchemists: it purified substances, yielding clues about the true nature of matter—and it did so in a manner reminiscent, at least, of the way a scientist might work. The history of the still, the other alchemical instruments, and of the practical procedures for their use in manipulating chemical compounds is a story of the alchemical commitment to a tradition of empirical learning. Alchemists used their instruments to muck about with real stuff, to perform what were in fact experiments to identify how different chemical compounds formed, altered, and could be transformed from one state into another.

But if Newton's alchemy could at times behave like what we could call science, what of the philosopher's stone? The "true aqua vitae"? And what of references in Newton's notebooks to "the broth of Medea" or "the horn of Amalthea"? Sir Isaac Newton is first and most the symbol of the rational in the study of nature; the architect of the method of science that "feigns no hypotheses" and constructs natural laws within the rigorous form of mathematical argument. Who is this other man who pored over ancient texts; who pottered about in a shed for weeks on end, heating, cooling, distilling, searching for the secret of manufacturing gold; who sought "noble" powers, too much like God's to be shared with any prying eyes? To historians today, the existence of Newton the scientist and Newton the magician in the same being has seemed a paradox, at best. John Maynard Keynes, the first modern writer to address Newton's interest in alchemy, did not even attempt to explain it: "Geniuses are very peculiar," he concluded. And yet Newton himself saw no contradiction in his interests, and his contemporaries, whatever else they thought of him, did not call him strange.

To understand the significance of Newton's passion for alchemy, one assumption must be made: Newton was a scientist; whatever he attempted, therefore, was part of what he understood the task of science to be. Modern science as Newton and his contemporaries and successors built it turned upon the discovery of the underlying order that governed nature, expressed in the abstract and general terms of mathematical laws. That sense of science derived, as we have seen, from a conception of patterns in nature and the notion of natural law that grew from its origins in ancient Greek science. It relies on a top-down approach to nature, one that searches for the unifying concepts from which every observable phenomenon could be derived. But there is a bottom-up approach as well, one that attempts to understand how the differences seen in nature came to be. Alchemy asked what the world was made of and what governed the continuous process of transformation in nature. In answering that question, the alchemist focused on the particular: how each combination of elements interacted; what each chemical process would produce; what incantations, what astrological factors, what subtle variations in the art of the adept could succeed or fail in the creation of alchemical truth. The clues lay hidden in the details, without which, the magus believed, all the knowledge of grand patterns would produce no more than pretty pictures, too abstract to be of use.

In their origins, these two traditions were distinct, even antagonistic. Mathematics was high culture; magic definitely low. What Newton recognized, though, was that the creation of a comprehensive picture of nature required a synthesis of both. The claim made by those who followed Newton, that their revolutionary science could account for every material phenomenon, implicitly assumed that modern science could travel equally well from the least detail to the whole picture as it could the other way around. Before that assumption could be made, before the scientists of

"The Oratory and the Laboratory"—musical instruments alongside a still and other alchemical apparatus—both sets of machine tools for exploring the harmonies of nature.

the eighteenth and nineteenth centuries could embark on their dream of reason, their predecessors, Newton prominent among them, had to tame alchemy, had to transform what had been magic into what could now be recognized as science.

From the very beginnings of the endeavor, of course, the search for the true nature of matter and of change was, if not a science, at least a rigorously defined art. The European version of alchemy emerged in the melting pot of Greek and Roman-ruled Egypt, in the first centuries of the Christian era. The creation of one material out of another was taken as a matter of course. The fifth-century thinker Aineas of Gaza wrote: "The changing of matter for the better has nothing incredible in it. Thus it is that those learned in the art of matter take silver and tin, make their externals disappear, color and change the matter into excellent gold. With divided sand and soluble natron [soda, perhaps, or more generally, any white salt] glass is made: that is a new and shining thing." Such novelty served a higher purpose—Aineas also wrote that his aim was to transform "the perishable and corruptible . . . which by the creative art becomes pure and beautiful"—but the early alchemists clearly perceived that their path to such spiritual gains lay with the mastery of the physical processes that could produce "a new and shining thing."

To do so, the alchemist sought to work upward through the thickets of all the malleable details of nature. Alchemical philosophy held that the substance of nature arranged itself into a hierarchy. All matter had some trace of the incorruptible purity that was the ultimate goal. Gold alone was truly pure, a blend of the four essences of earth, air, fire, and water in perfect balance—proved by the fact that gold does not tarnish, discolor or rust. All the other metals, and by extension all of nature, were created out of an unbalanced mixture of qualities. Each was composed out of a different arrangement of earth, air, fire, and water, yielding a particular accidental form. Some metals were better than others—silver was very close to gold, but lead and tin, for example, were "base" metals, far removed from the perfect balance of gold.

But despite the distance of the basest metals from gold, the existence of a continuous hierarchy suggested a practical strategy that could lead to the transmutation of metals. The alchemist's job was to find the procedures that could alter a particular compound to emphasize at least one pure quality—of color, malleability, incorruptibility, and so on—that could by a series of such operations produce perfect, immortal gold out of the least promising mixtures of ordinary metal. The earliest, most primitive methods attempted by the alchemists emerged directly out of the craft traditions of metallurgy and practical chemistry that Egyptian and Greek artisans had developed, innocent of any abstract philosophizing. Egyptian papyri dating from the third century of the common era record recipes that emphasize the coloring of mixtures of metal to create alloys that resemble gold or silver. One calls for the heating of gold until it is bright, and then adding enough of two coloring compounds, "misy" (most probably a natural mixture of iron and copper sulfates) and the other a reddening dye, to equal the amount of gold—doubling the amount of gold-seeming metal. Another reads: "Add 6 parts purified tin and 7 parts Galatian copper to 4 parts silver, and the resulting product will pass unnoticed for silver bullion."

The anonymous scoundrel who wrote that had his ends clearly in view; certainly, officialdom saw alchemy in much the same light as counterfeiting. There are reports of attempts to suppress alchemy, the most famous being the story Gibbon repeats of the emperor Diocletian's order to burn of alchemical texts, to block any attempt to finance a revolt against Rome with manufactured treasure. This attitude certainly encouraged secrecy, but it did not stop students of the art from pursuing the possibilities suggested by their experience in smelting and forging metallic ores. One broad approach to the problem of transmutation sought to mimic what was understood to be the process that generated gold in the wild. The theory was that metals grow just like living tissue, only more slowly, changing their form and nature as they proceed through different developmental stages. So the alchemist attempted to "grow" gold from a "seed" planted in ground that had already been prepared. The alchemist would begin, following this strategy, with some mixture of the four base metals—iron, tin, copper, and lead—melting them into an alloy that turned black when exposed to air. That blackness, seen as the absence of color, was the starting point from which the entire process of transformation could begin. This blackened lump would next be whitened, raising the sample to the status of silver. Some alchemical recipes suggest adding a drop of actual silver at this stage, as the catalyst that could foster the conversion of the entire body (an idea which certainly lent substance to the accusation of counterfeiting, or at least fraud). But the actual transformation was supposed to occur when the black metal, together with its leaven of silver, was melted again with mercury, to produce a lump of metal with a bright, shining, white surface sheen. This was the alchemist's silver—not to be confused with the ordinary metal—white on the outside but yellow inside. With the addition of a seed of gold and a coloring agent, often sulfur, the metal would be melted a third time. The product would then finally emerge, colored yellow both within the lump of metal and on its surface: gold, made by human art. A fourth stage, to produce a violet-colored metal seen as far superior to common gold, could follow, depending on the aims of the alchemist.

Other recipes pursued different chemical pathways to the same end, the production of a lump of metal possessing the color and physical appearance of gold. The proportions and the number of steps varied—one famous procedure used copper, sulfur, lead, arsenic, and other materials, but left out iron, tin, and mercury, to produce an alloy to be transformed by contact with true gold—but the underlying idea was the same: the production of gold was an ordinary, replicable process that depended on the discovery of the relationships between one form of matter and another. The alchemist's ability to control each step of the process turned on his understanding of those connections. The need for that control provided the impetus that led to the creation of some of the fundamental instruments of applied chemistry; in one form or another, they have remained in continuous use ever since.

The person most often associated with the invention of the laboratory apparatus of alchemy is Maria the Jewess. Possibly a legendary figure, none of her writings have survived intact, but the ideas attributed to her were preserved as quotations in other alchemists' work. She most probably flourished in the fourth century, and as she developed her own method of transforming substances (not just gold), she built a variety of instruments. She used several different methods of heating her preparations: besides the furnace with its direct flame, she used the slow heat produced by the fermentation that takes place in a bed of dung; the more rapid, even heating that can be generated in a hot ash bed; and the gentle transfer of heat that takes place within a bath of water kept just short of the boil. (This last survives, its alchemical heritage forgotten, as the bain-marie—Mary's bath—a kind of early steam table.)

The most important contribution attributed to Maria was the refinement of a device that could extract the pure essence of some undifferentiated substance: the still. Before Maria, some observers had noticed that drops of liquid could condense on the lid of a pot in which something was being boiled, and that the condensate on the lid could differ from whatever was being heated beneath it. The distillation of seawater to produce fresh water out of the condensing steam was well known in the ancient world—commentators on Aristotle, if not the master himself, knew the procedure. But later generations gave the admittedly shadowy figure of Maria the credit for taking the hints that suggested the process of distillation and to have laid out the first clear design for a machine specifically intended to produce and draw off chemical distillates.

That still was actually a remarkably sophisticated instrument, designed by someone with a clear, technically adept mind. Maria provided quite precise details for its construction: one text quotes her calling for "three tubes of ductile copper a little thicker than that of a pastrycook's copper frying pan." The whole still consisted of a bottom chamber, in which the material to be distilled would be heated; a long tube through which the steam could rise; an upper condensing chamber, pierced by the three pastry-inspired pipes leading downward to three collecting vessels that could capture the distilled fractions.

This still incorporated two major advances over the earlier, embryonic designs. Condensing gas releases latent heat as it makes the transition from the vaporous to the liquid state, warming the condensing chamber of a still. A still's efficiency depends on keeping that chamber cool enough to move the condensation process along briskly. Sticking the condenser well above the boiler, in an exposed enough position to permit the alchemist to cool the chamber with air or flowing water, increased both the quantity and quality of purified distillate. The improvement in purity came about because different elements within a mixed source will boil off at different times and temperatures. If the condenser is too slow, different vapors mix and blend in the distillate; if condensation occurs rapidly, increasingly pure fractions can be drawn off. The other innovation ascribed to Maria, the use of three pipes penetrating the condenser at different heights, further improved her ability to separate the components of her still's output. Different materials have different condensing temperatures as well as different boiling points—which in a still translates into condensation at different heights of the condensing column. (The farther up they go, the longer the time the gases have to cool and condense.) Pipes leading off from the still head at different heights will harvest different mixtures of condensed liquid and still-gaseous vapor, producing distillates of greater or lesser purity. (If the distillates don't mix as liquids—like oil and water—pipes at different heights can be used to draw off the water in one direction while siphoning away the oil layer that would rest on top of the water, similarly separating and purifying the products of the distilling process.)

The still thus vastly increased the ability of the early alchemists to explore the detailed behavior of matter. Distillates were themselves clearly transformed substances—purer, better than the original mixtures of base matter. The "divine water" produced in the still could foster the further transformation of metals; Maria used her still to blend copper and lead with vapors, often of sulfur compounds, to produce black sulfides with which to start the transmutation process. She boasted that her prime matter was superior to others, in fact, because metal fused with sulfur is black all the way through rather than simply black on the oxidized surface of an ingot exposed to air. She also experimented with sulfur-mercury compounds, and later alchemists credit her—or at least the use of her still—with producing the deep red tinctures of mercury sulfide that served to transform whitened alloys into something resembling gold.

Maria herself, as much as she can be glimpsed in the fragments she left behind, clearly reveled in her mastery of alchemical technique. Her design addressed and solved the major problems blocking the development of an efficient still, and she explored at least somewhat systematically its use in producing distinctive chemical reactions, alloys, and new compounds. But to express her accomplishments in the language of modern chemistry obscures Maria's most important role. For her the still was not merely a piece of apparatus—rather, the machine expressed a fundamental truth about the relationship between any human craft and the universal processes of change in both the natural world and the world of the spirit.

The still, that is, contained the universe in microcosm. The Arab alchemist Ibn Umail, writing in the tenth century, restated Maria's picture of the still in action:

> Maria also said: The Water [the distillate] which I have mentioned is an angel and descends from the sky and the earth accepts it. . . . She meant by this the Divine Water which is the Soul. She named it Angel because it is spiritual and because that Water has risen from the earth to the sky of the still. As for her statement [that the Water] descends from the sky, she meant by this its return to the earth; and this Angel she mentioned I shall explain to you another way. . . . She meant by this the Child which they said will be born for them in the Air while conception has taken place in the Lower—this being through the Higher Celestial Strength which the Water has gained by its absorption of the Air.

Read across the divide of time and language, alchemical writings are genuinely obscure. The penchant or necessity for secrecy encouraged the use of codes and hidden messages. Even a writer like Ibn Umail, trying to explain Maria's method, uses terminology whose meaning has shifted or been forgotten. It is tempting to read metaphors into Maria's theory—to substitute the idea of vapor for Angel, condenser for sky—and thus to recast

the account as a poetic expression of a familiar, modern understanding of chemical processes. Doing so utterly misses the point.
Alchemy was never simply a technique. Rather, it persisted, and
alchemical ideas flourished in such disparate cultures as those of
Greece, Islam, India, and China, because (at least partly) it persuaded some of the most skilled people in those cultures that its
methods could in fact reveal fundamental truths about the universe as a whole that were of extraordinary value to humankind.
Maria and those who followed her had a coherent, if foreign,
theory of the world, one they at times tried to state as openly as
possible. The trick of understanding their thought is not to replace their words with our own but to assume that the alchemists
said what they meant.

To Maria, then, the still was a machine, making her Divine
Water, but it was much more besides. It functioned exactly as
nature did, creating vapor, clouds, rain, and sustenance; it operated under the natural laws of generation and growth—the new
product, the distillate, was conceived and born; finally, the still
acted in the realm of the divine: the risen spirit (a material, real
thing) was infused with power as it partook of the connection to
the angels and the "higher celestial strength" that permeates all
of nature and all human experience. These were not mere analogies. The "work" was the key—the series of transformations,
the symbolically and actually significant alterations of the colors
of metals that could ultimately reveal the "true" character of a
substance, its noble status as a pure, refined essence of gold. The
still in which such work took place was so useful, alchemical
operations in general so numinous, precisely because they replicated processes in nature, offering alchemists the chance to manipulate what went on in the macrocosmic universe through their
control of the microcosmic realm.

In the European tradition, the crucial idea of the unity of the
microcosm with the macrocosm, the entire universe, is most
deeply linked with the figure who became for Renaissance
and early modern Europe the patron saint of alchemy, the
philosopher-god Hermes Trimegestus—Hermes the Thrice
Great. Hermes is associated with both the Greek god of the same
name and with the Egyptian god Thoth. He received the alchemical revelation: "Hermes saw the totality of things. Having
seen, he understood . . . what he knew, he wrote down." He is
supposed to have written the basic texts on transmutation and to
have left the fundamental credo that animated alchemical re-

search. That statement, allegedly found on an emerald tablet clutched in Hermes's hands at the time of his death, reads in part: "What is below is as what is above, and what is above is as what is below. . . . Separate the Earth from the Fire, the Subtle from the Gross, smoothly and with judgment." Maria with her still was attempting to do just that: to identify on the small scale the critical steps by which nature at large produced one substance out of another. Gold was a by-product—an assay, as it were; proof, when it was produced, that the alchemist had got it right. The real prize was power over nature and the knowledge of those essential truths that pervade the world, wisdom the ability to make gold would merely affirm.

The Greco-Egyptian alchemists of the fourth and fifth centuries would go on to develop the basic idea. Kleopatra, following Maria, produced an even more elaborate theory than her mistress's of the natural cycles of birth, growth, and decay that govern (among all else) the production of metals in nature and in the alchemist's workshop. Zosimos, working around 300, shared Maria's fascination with apparatus, wrote about furnaces and recalled the engineering ideas developed by Archimedes. He also studied the process of transformation and saw the still as a womb in which birth, transformation, and rebirth took place as a metal gave off its spirit as vapor and reemerged as a newly formed, altered body at the end of the distilling process. And on the practical side, the first glimmerings emerged of the concept of the philosopher's stone—a substance or a method that could turn all it touches to gold. The search for the stone triggered epidemics of gold fever among medieval and Renaissance alchemists, who sought its secret in ancient texts. But what became a mythical object evolved out of alchemical processes like those that produced the sulfides of mercury that could tint vast quanities of whitened metal with a color that resembled gold.

These ideas survived the catastrophe that swept away the culture in which Western alchemy was formed. The Egyptian melting pot in which Greek philosophy, the Near Eastern craft tradition, and the Alexandrian heritage of empirical research could meet and mate was overturned in 638, when the first rush of the Islamic conquest swept away the last vestiges of Roman rule in Africa. Islamic scientists, however, proved to be as fascinated by alchemy as their (vaguely) Christian predecessors had been. Jabir ibn Hayyan, or Geber, as medieval scribes wrote his name, was the most famous name associated with the Islamic pursuit of alchemical knowledge, flourishing in the latter half of the eighth century. He promoted the idea that all metals are made of some combination of sulfur and mercury, the elements of fire and water, which mix in different proportions and purities to produce all the observed variety in nature. The sulfur-mercury theory set up the fixed target that alchemists could try to hit, for gold was defined as a compound of the two in proportions that would strike a perfect equilibrium between their fiery and watery impulses. (Jabir apparently tested this idea and concluded that the ordinary forms of sulfur and mercury found in nature were in fact only the closest accessible relatives to some still-undiscovered perfect elements that would display the properties called for in his theory. If that kind of stretch sounds outlandish, consider one of the criticisms leveled at Darwin when he proposed his theory of the origin of species, namely, that the earth was believed to be too young for evolution to have taken place as he suggested: too little time had passed to produce all those different organisms. Darwin's only answer was that someone would eventually prove that the earth was much older than then thought—which is exactly what happened in the early years of the twentieth century.)

Back in the eighth century, Jabir was in fact an accomplished practical chemist: he used a still to extract acetic acid from vinegar, he studied the production of dyes and inks, worked in metallurgy, and he discovered nitric acid. (The precise record of his accomplishments is hard to fix, because many of the works attributed to Jabir probably were written well after his death. His reputation was such that he became something of an Arab Hermes, a figure at least partly legendary from whom anonymous authors hoped to gain authority for their ideas.) The historical Jabir lived into his nineties, dying in internal exile sometime in the early 800s, having backed one of the wrong factions against the caliph Harun al-Rashid (of *Thousand and One Nights* fame). During the great flood of cultural interchange in the twelfth and thirteenth centuries, Christian Europe would receive Jabir's (or Jabirean) ideas along with those of the Greeks and Egyptians whose thoughts the Islamic scholars had preserved. The alchemical tradition transmitted to the West contained the twin themes that dominated Jabir's work. One clear emphasis came in Jabir's empirical pursuit of useful results. He wrote: "The first essential in chemistry is that thou shouldest perform practical work and conduct experiments, for he who performs not practical work nor makes experiments will never attain to the least degree of mastery. . . . Scientists delight not in abundance of material; they rejoice only in the excellence of their experimental methods."

The other preoccupation that Western thinkers came to share was the refined, Jabirean version distilled out of the older theories of alchemical cosmology. The mercury-sulfur theory of metal and its transmutation retained the essence of the original philosophy: first, that the entire cosmos was formed in all of its magnificent variety out of the same stuff, shaped by the universal processes of growth and decay; next, that any single piece of the whole, any microcosm, would display the same elements and processes that governed the whole, the macrocosm; and finally that the ability to manipulate affairs on the microcosmic level would expose the processes that shape events in nature at large. As alchemy progressed in the West, these ideas provided the motive, while the genuine successes of alchemists acting as chemists suggested the means with which to move forward.

One practical advance came quite swiftly, in an area the Arabs had ignored. The medical school at Salerno, founded in the ninth century, lay at a crossroads, with access to Greek, Norman, and Arabic ideas. Someone at the medical school picked up on one of the uses Arab alchemists had devised for the still, extracting and purifying compounds to be used as drugs. One advantage the Christian researchers had over their Arab counterparts was somewhat greater freedom of action: fermented liquids were forbidden to the Islamic faithful, but an unknown physician at Salerno apparently hit upon the idea of placing wine in the boiler of a still and extracting its spirit.

The trick in producing distilled liquor lay exactly where Jabir placed his emphasis—on the excellence of the experimental method. The purification of alcohol in a still depends on the ability to cool the still head sufficiently, as alcohol has a lower condensing temperature than water. Medieval stills had condensers well removed from the direct heat that fired the boiling chamber—evolving into designs that could produce spirits in bulk. The earliest concoctions were rare, expensive, and probably used for purely medicinal purposes (as good an excuse then as now). Strong drink became generally popular after the plague epidemics of the fourteenth century, when the legend spread that those who drank regularly would not die of the disease. It was known by a number of names—aqua vitae (the water of life) whiskey, burnt wine, and so on—but the term *alcohol*, just like the spirit itself, has its own, distinct alchemical heritage. Alcohol, *al-kuhl* in Arabic, was the name of the black paint used as eye-lining cosmetics in the advanced capitals of the East. The development of cosmetic colorings was itself a sister craft to alchemical investigations, and the term slowly gained a more general meaning— first, it referred to any finely divided powder, but ultimately it gained the sense of the "finest part" of something. Paracelsus, the famous sixteenth-century alchemist, doctor, and drunk, apparently felt that the strong spirit the still produced was the finest part of wine—and hence gave the stuff its name: alcohol of wine, or simply alcohol.

The invention of distilled liquor was an early success that turned on the use of alchemical methods—but as whiskey making became an industry, rather than a specialized branch of medicine, much of the interest in the ideas of alchemy turned elsewhere. The great European revival of alchemy as both an empirical investigation and as an expression of a broader philosophy took place during the Renaissance, when successive generations of thinkers sought to replace earlier ideas about nature with their own. The interest in gold making remained a centerpiece of alchemical research—one text printed in 1535 offered five recipes for making gold and noted that while transmutation was difficult, the discovery of metals unknown in classical times made the job easier.

Such claims did not go unchallenged. One of the pleasures of the Renaissance is that its makers were contentious, proud, and plain-spoken people. Leonardo da Vinci, for one, had a bitter contempt for what he saw as sloppy thinking and took dead aim on alchemy: No alchemist, he said, "either by chance or deliberate experiment succeeded in creating the smallest thing which can be created by nature." (He did believe, however, in the traditional theory that gold in nature grew by a slow process of transmutation, changing whatever the original seed metal touched within the earth.) But Leonardo's objection to alchemy was conditional, limited: he scorned the discipline, and others like it, because its experiments did not seem to work, but remained open to the possibility of some alchemical success.

And da Vinci essentially ignored the alchemical theory of nature—which is what some Renaissance men sought to put to their own, novel use. What they made with it they termed a new kind of magic, natural magic, intended to be quite distinct from medieval superstition and the black arts of necromancy. In the seventeenth century the creation of a new science of nature would demand the demolition of magic and of the tradition of the magician, the magus. But before that new broom could sweep through, the natural magician had his say, and he bequeathed to his scientist successor critical elements of the scientific style of thought.

From the outset, natural magic, as its devotees defined it, elevated the alchemists' concern for the particulars of the material world to a central goal of any investigation of nature. As they did so, alchemists found themselves locked in the pursuit of the causal mechanisms that could connect each event with the next in the endless unfolding of change in nature. The job at hand, was to connect each detail of human experience to the large-scale organization of the cosmos. To do so, the new magi tried to build a systematic depiction of the links that bound one phenomenon to the next. They argued that the *spiritus mundi* (the spirit of the world) penetrates "the whole of the sensible universe," as the visionary Marsilio Ficino wrote in 1489. Ficino went on to write: "It vivifies everything, everywhere, and is the cause of all generation and motion," giving a new twist to the old idea of unity between microcosm and macrocosm. Maria had thought to imitate natural processes with her still. The magi left behind much of the purely alchemical interest in apparatus—Ficino sought a more direct connection to the world spirit. Where the alchemist would use her still to re-create natural processes, Ficino developed an elaborate, detailed, complex set of symbols and images to represent all of the components of the observable universe. As he understood it, the correspondences between the symbols and the symbolized, when drawn correctly, would break down the gap between the two—the pictures could become what they pictured. When they did, Ficino's description of nature would become a kind of natural process in itself, capable of affecting the course of events out in the world of experience.

So Ficino made drawings, images, talismans. The magus could gain long life (for himself, and presumably for someone else) by turning to Saturn, the planet that governs longevity, and making an image to precise specifications: on a sapphire, engrave the picture of "an old man sitting on a high throne or on a dragon, with a hood of dark linen on his head, raising his hand above his head, holding a sickle or a fish, clothed in a dark robe." If, having followed the instructions above, you wanted your long life to be

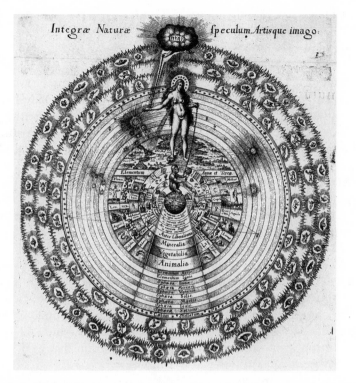

Another of Robert Fludd's representations of the unity of the cosmos, from earth, through all the elements, the planets, and the stars.

happy, you could also make an image of Jupiter on a clear stone, with similar attention to detail.

Ficino told of talismans that could cure diseases; images that could improve their bearers' wit; devices to gain wisdom. His account of these supremely useful instruments is straightforward, matter of fact, as if his magic were simply one more facet of daily experience—which for Ficino, and for many others of his day, it was. Ficino argued that the efficacy of his talismans turned on what were self-evidently natural processes. Each of his images derived its potency by drawing down from the heavens the spirit that animates each of the planets—in effect, becoming that planet. An icon of Venus, clad in white and yellow, holding flowers and apples, Ficino promised, brought the distant influence of the cosmic object directly into the grasp of the person who peered at her image—holding out the prospect of happiness and strength. In his grandest ambition, Ficino asked: "Why, then, should we not permit ourselves a universal image, that is, an image of the universe itself? From which it might be hoped to obtain much benefit from the universe."

The key to Ficino lies with that hope: his magic, good magic, natural magic, functioned precisely because it used an idea of the universe to gain advantage within the universe. It avoided any supernatural tricks. He had no truck with demons, none with the black arts. Natural magic was simply the systematic use of natural phenomena—like the known, potent influence of celestial objects—to achieve desired ends on earth. His initial assumptions are (and were, to some of his contemporaries) not without their problems, but once they are granted, Ficino's elated perception of a deep understanding of nature makes sense. So does, from his vantage point, the belief that he had mastered the seemingly arbitrary forces of change within nature that so vex human affairs. In a letter to Lorenzo di Pierfrancesco de' Medici, Ficino promised the young prince: "If you thus dispose the heavenly signs and your gifts in this way, you will escape the threats of fortune." A proper understanding of the web of natural connections that bind together all of the sensible universe meant, Ficino said, that the person who conducts himself correctly within that web "under divine favor will live happy and free from cares."

Magic to the magus, that is, was merely an inquiry into nature, just like any science, with its own methods, tools, and categories of thought. Ficino's contemporary and fellow magician Giovanni Pico della Mirandola simply asserted that "magic is the practical side of natural science" and proceeded to lay out the clearest expression in Renaissance thought of the role of magic within the scientific enterprise. Pico cut a romantic figure in his day: aristocratic, arrogant, and possessed of enough dramatic sense to die young. He broke onto the public scene in 1486 as a twenty-four-year-old prodigy, publishing in Rome nine hundred theses on philosophy, magic, and other arts. He offered to pay travel expenses from anywhere in Italy for anyone who wished to dispute him, but the debate never took place. Instead, Pope Innocent VIII appointed a committee to investigate the whiff of heresy in Pico's claims. The committee found thirteen heretical propositions, and Pico made his position worse by trying to defend them in print—which led to his abrupt departure from Rome and several years of dodging papal authority. Lost in the hullabaloo was a short text that Pico apparently meant to deliver as the introduction to the disputation. Titled "An Oration on the Dignity of

Man," it appeared only after Pico died in 1494, just thirty-one years old. Against the background of his short and very public life, the "Oration" became what the historian Frances Yates called "the great charter of Renaissance Magic."

Pico first drew the essential distinction between the black arts and appropriate magic: "Magic has two forms," he wrote, "one of which depends entirely on the work and authority of demons, a thing to be abhorred. . . . The other, when it is rightly pursued, is nothing else than the perfection of natural philosophy." The true magician, Pico claimed, rejecting "the rites of evil spirits . . . embraces the deepest contemplation of the most secret things, and at last the knowledge of all of nature." The magus, "in calling forth into the light as if from their hiding places the powers scattered and sown in the world by the loving kindness of God, does not so much work wonders as diligently serve a wonder-working nature." Pico's magic, like Ficino's, is simply the expression of natural law. His magician, "having searchingly examined into the harmony of the universe . . . and having clearly perceived the reciprocal affinity of natures, and applying to each single thing the suitable and peculiar inducements, brings forth into the open the miracles concealed in the recesses of the earth . . . and as the farmer weds his elms to vines, even so does the magus wed earth to heaven, that is, he weds lower things to the endowments and power of higher things."

Pico called this craft a science—and with his definition of natural magic he made explicit how his magic takes on the character of scientific investigation. First, like Ficino, Pico reaffirmed that every detail of nature matters. The magus's power depended on his understanding of the harmony of nature—how it all fit together. Pico's magus must pay attention to "each single thing." Without accounting for each of nature's parts, the magus could not coax out the secrets hidden within the welter of natural phenomena that generates the miracles to which Pico laid claim. From this stand, Pico next took one giant step forward. Individual phenomena, he claimed, took place within nature, but a nature newly conceived: nature as an actor, an agent. Its forces of generation, growth, and change animate the natural world, creating a living force. Nature—not God—worked the miracles, and it is in the material universe in which human beings dwell, and to which human beings have direct access, that the secrets of such mysteries wait to be discovered.

Pico had a capacious view of what that universe might contain.

He even identified bad angels—Azazeal, for one—who lay in wait to devour the magician who erred in his incantations and stumbled across a diabolic lair. But his conception of magic as the precise control of events taking place within the material world turned on the central belief that all effects have their causes—natural causes—which the human mind can uncover and understand. The logic that culminated in the charms of natural magic led directly to a larger sense of the same thought: that a chain of cause and effect could be found to account for any phenomenon that a human being could observe or experience. The older magic that invoked the special powers of the supernatural—or of that God that could suspend natural law—inhabited a world in which human affairs could be manipulated in ways no one could necessarily predict or prescribe. The natural magician, by contrast, asserted that even God scattered and sowed his powers through nature, as a gift to humankind. The scientist, like the magician (or as Pico would have put it, one man both scientist and magus), was thus compelled to turn to nature itself, to engage directly the mysteries that lay before him.

As expressed by natural magicians, this scientific sensibility set the standard for ways of thinking about nature well before the scientific revolution coalesced. Evidence for the shift turned up in seemingly the least likely of places. The sixteenth and seventeenth centuries were bedeviled by witches who troubled the faithful and by devils who could overwhelm the soul and steal a human body for their own ends. Both the Catholic and the newer Protestant churches developed elaborate techniques of exorcism—and in keeping with the emerging idea of natural causation, they constructed what they saw as the science of demonology.

To begin with, exorcists clearly defined possible causes and characteristic symptoms of what might appear to be possession. The three mechanisms as laid out in a French case in 1599 were "sicknese, Counterfeiting, or Diabolicall possession." The characteristic symptoms that would betray the presence of a devil were the ability to understand or speak foreign languages, inexplicable knowledge of secrets, abnormal bodily strength, and terror in the face of holy objects, Scripture and the like. The historian D. P. Walker published an account of that case in his study of exorcism, *Unclean Spirits*, reporting on an attempt to confront the devil with the tools of reason. It began when a twenty-six-year-old woman named Marthe Brossier appeared in Paris in the beginning of March 1599 claiming to be possessed by Beelzebub himself. On arrival in the capital, Brossier manifested some of the classic attributes of a demoniac presence: she had fits, went into convulsions, and railed against the Huguenots, asserting that for Beelzebub's part, all such Protestants belonged to him. That sufficiently inflamed Catholic-Protestant tension at a time when the once Huguenot, then Catholic King Henry IV of France was attempting to calm the situation, so that Brossier was subjected to formal study by a team of physicians appointed by the bishop of Paris. The assembled doctors examined Brossier and rejected the possibility of disease: the two organic causes proposed were epilepsy and hysteria. As epilepsy was understood always to be accompanied by a loss of consciousness, and hysteria by loss of breath, neither of which Brossier experienced, the physicians looked to the other two possible mechanisms. Here they split ranks. Those experts disposed toward a diagnosis of diabolic possession cited Brossier's apparent ability to understand foreign languages, as she responded, more or less, to questions put to her in Greek and in English, and to a hint that she might be clairvoyant. In addition, they pointed out that Brossier did not feel pain during her fits even when stuck by a pin, and that she showed no signs of exertion or reaction after each of her bouts with Beelzebub.

Throughout their account, this group of physicians attempted to display careful attention to the causal mechanisms of apparent possession, employing empirical tests, as when they pricked Brossier. They supported their conclusions within a framework of formal logic: given three competing models of the disorder before them, they sought falsifying tests that their subject could pass or fail to isolate the single true cause of the phenomenon before them. Unfortunately for the elegance of this approach, as Walker tells the story, a rival team of doctors had also examined Brossier. Using the same methodology, and considerably more rigor, they came to an opposite conclusion. One of them, the physician Michel Marescot, prepared a damning exposé of what seemed to him a clear fraud. He found that Brossier did feel pain, that her fits were not terribly violent, that she guessed the meaning of questions in Greek and answered vaguely enough to get by. His case was bolstered by the fact that before Brossier arrived in Paris, the bishop of Angers had examined her and found that her Beelzebub failed the basic test of demoniac gastronomy:

instead of going into convulsions on receiving a draught of holy water, Brossier lay still, and when offered ordinary water that she had been told was consecrated, she threw a fit.

Marescot concluded that the case had "nothing from the Spirit, much counterfeited, a few things from disease"—that Brossier was a fraud, and possibly sick as well. He diagnosed Brossier as a melancholic, prey to self-delusion. Finally, Marescot issued a general warning against recklessly blaming the devil for unusual events—referring his readers to texts on natural magic for examples of phenomena both rare and purely natural.

In the Brossier case, as in other famous tests of demoniac possession, both those who doubted and those who believed in the presence of the devil granted the possibility of diabolic influence—but both sides accepted the requirement to test for the devil's spoor using language and methods suitable for any inquiry into a natural process. In doing so, the scientists of witchcraft went beyond Marescot's caution that unusual events might not involve the devil. Even when the devil lays his hands on human affairs, they insisted, he, too, must be constrained by the natural order. As the English demonologist John Cotta put it in 1616: "Though the divel indeed, as a Spirit, may do and doth many things above and beyond the course of some particular natures: yet doth hee not, nor is able to rule or commaund over generall nature, or infringe or alter her inviolable decrees . . . neither is he generally Master of universall Nature, but Nature Master and Commaunder of him." The devil could wreak havoc through natural processes—causing fits or convulsions that resemble those of epilepsy, for example—but he could not overturn the universal laws of nature.

The study of demoniac phenomena, thus, was simply the examination of natural events for signs that the devil had intervened. It was not without its methodological difficulties. Demonologists pointed out that though the devil could not overturn natural law, the Great Deceiver could nonetheless persuade the imperfect senses of human observers that the impossible had occurred. Nonetheless, the careful investigator was on the alert to penetrate any obscuring tricks, at which point, as Martin Biermann wrote in 1590 in his guide to magic and witchcraft, "magical actions and motions are reducible to the considerations of physics."

This cool voice of reason would sound even more modern than it does were it not for the context. During the Renaissance, occult studies—demonology, natural magic, alchemy—took on some of the trappings of science. In doing so, those disciplines brought the search for the links in the chain of causation to the front of any investigation of nature, which at least some investigators of the occult were able to put to good use. In the last great flurry of alchemy immediately preceding the scientific revolution, its practitioners established a major element of the foundation of modern chemistry. But every appeal to magical mechanisms behind natural processes, every judicious linking of the devil to the workings of natural laws, still left Lucifer in the game. For all their insight into chemical phenomena, the alchemists' final serious attempt to create a new synthesis of natural knowledge foundered on the ultimate failure of their occult theory of cause and effect.

At first, though, such theoretical quibbles didn't seem to matter that much. What counted was whether or not alchemical practice could produce anything of value—and the new, or newly revived, alchemy, associated with the development of natural magic, seemed to hold out (again) the possibility of limitless riches. In the early sixteenth century, Hans Fugger's descendants ran Antwerp, the financial engine of sixteenth-century trade. They had a substantial beachhead in Venice, astride the commercial routes leading east. They had an office in Chile and looked to expand farther, into Asia. Most of all, they had metal. The Fugger family controlled central Europe's most significant mining enterprises, holding sway through Hungary, Bohemia, the German Alps, and the Tyrol. At Schwaze, in the Tyrol, the Fuggers extracted copper and silver—and ran a mining school, led by Sigismund Fugger. That school transmitted the technical knowledge of metals the Fugger enterprises had accumulated over more than a century of mining, but Sigismund himself sought still deeper knowledge of metal, as he pursued a lifelong program of alchemical research.

That Fugger hunger for the common ores that could be extracted from the earth, along with the extraordinary metals produced by the refinement of the alchemical art, captures the commercial spirit of at least some Renaissance alchemists. From the Fugger perspective, of course, alchemy could be seen as simply a sound business interest—research and development—leading, perhaps, to ongoing control of a scarce resource. With vicious accuracy, the satirist Ben Jonson skewered the money lust of some would-be alchemical adepts in his play *The Alchemist*, first performed in 1610. In one scene, Mammon beseeches the

"The Alchemist and His Assistant"—a gently mocking view of the quest for gold.

alchemist Subtle to provide him with the philosopher's stone that could turn all his metal into gold. Subtle warns: "Why this is covetise!" and Mammon responds: "No, I assure you/I shall employ it in all pious uses,/Founding of colleges, and grammar schools,/Marrying young virgins, building hospitals/And now and then, a church."

But money was not the sole or even the primary goal of Sigismund Fugger's most famous student. In 1522, a young, reckless, already errant Swiss named Theophrastus Bombastus von Hohenheim came to Schwaze and enrolled in the Fugger academy. That man, better known as Philippus Aureolus Paracelsus, would stay with Sigismund Fugger for just one year—long enough to gain the foundation of what would become a radical new application of alchemical theory. Upon leaving Schwaze, Paracelsus wandered for three years, weaving through Germany, Italy, France, Russia, and beyond. He worked briefly as an army surgeon and acquired, somehow, a doctor of medicine degree. (No university claims him as an alumnus.) His mind was a sponge, and as he traveled he gleaned whatever he could from whomever he encountered on the road: gypsies, magicians, apothecaries, the whole traveling carnival. He reappeared in Switzerland in 1526 and set out to transform the practice of medicine.

His first case was that of the publisher Frobenius, lying seriously ill in Basel. Paracelsus managed to cure him, earning the gratitude of Frobenius's good friend Erasmus, along with an appointment as Basel's official physician. From that bully pulpit, Paracelsus made his view of traditional medicine clear. He publicly burned the works of Galen and Avicenna and lectured the local doctors: "O you hypocrites, who despise the truths taught you by a great physician, who is himelf instructed by Nature, and is a son of God himself! Come, then, and listen, impostors who prevail only by the authority of your high positions!"

Paracelsus's message was that the proper treatment of disease turned on the discovery and preparation of the medicines appropriate to each case. The traditional practice of cobbling up mixtures of herbs and roots, oils and balms, earned his contempt: "They think it suffices if like apothecaries they jumble a lot of things together and say *Fiat unguentum* (let it be a salve)." Rather, Paracelsus urged, true physicians should extract from natural substances the pure essences that could be shown to have a direct mechanism in the treatment of an illness. Thus, such healers spend their time "learning the steps of alchemy. These are distillation, solution, putrefaction, extraction, calcination, reverberation, sublimation, fixation, separation, reduction, coagulation, tinction, and so on." Paracelsus's acute realization here was that diseases could be understood as specific phenomena to be analyzed in detail. Given that analysis, the alchemically sophisticated physician could then isolate the precise compound that would serve to restore the balance of the diseased body of the suffering patient.

As a matter of daily practice, Paracelsan ideas led chemists to focus on the search for biomedically useful preparations with an intensity and clarity of purpose never before achieved. By Paracelsus's time, the mid-sixteenth century, alchemists and their sometime allies—the metal workers, dyers, soap makers and other craft adepts of the sort Paracelsus himself consulted—had already accumulated and transmitted an enormous body of hands-on knowledge of chemical reactions. As it had for alcohol manufacture, the development of originally alchemical processes aided the dye industry, ink making, smelting, and metallurgy. Distillation remained a favorite technique, generating the discovery and then the commercial production of acids like the oil of vitriol (sulfuric acid—H_2SO_4), and aqua fortis (HNO_3). Paracelsus himself, restlessly wandering from place to place, collected an extraordinary catalogue of natural phenomena, listing not only the varying symptoms of diseases but (harking back to his youth) the experience of miners with different metals, tallies of novel plants, examples of the variety of craft techniques in town after town. But most important, Paracelsus substantially redirected the course of alchemical research: chemistry in both the ancient and modern sense first entered the medical curriculum as a result of his relentless campaign. In spreading his gospel, Paracelsus expressed one of the founding ideas of modern medicine: that drugs could be specifically targeted to unhinge the causal mechanism of a particular disease.

Yet for all his vehemence and the violence of his rhetoric (one opponent was damned as a "wormy and lousy Sophist"), Paracelsus himself produced almost no effective treatments of his own. It is hard to trace the details of much of his medical thinking, for though he wrote prodigiously, his works are convoluted, obscure, and difficult. Part of the problem lay with his love for alcohol—the Paracelsan legend has the master writing furiously, swiftly, into the small hours of the night, propelled forward by his skinful of drink. But the deeper problem lay with his underlying model of disease, his pathology. His picture of how the human body and its frailties was a kind of intellectual distillate of the entire accumulated mass of hermetic writing, alchemical texts, the teachings of natural magic. In Paracelsus, that tradition became perhaps the most clearly expressed alchemical statement of the connection between the human body and the universe. The starkness of that theory exposed its most vulnerable component—its conception of the causes of disease, of the way change occurs in nature—to the test of the same kind of empirical inquiry Paracelsus and the alchemists championed.

For Paracelsus, the microcosm-macrocosm connection was absolute. The human body in all its parts contained a complete representation of the larger universe, as Paracelsus interpreted literally the original expression of Hermes Trimegestus: "The macrocosm contains creatures of earth and water; man has fleas, lice and intestinal worms. The macrocosm has rivers, springs, seas; man has entrails. . . . The macrocosm contains exhalations that burst out in its bosom, for example the winds; man has his flatulences. The macrocosm has sun and moon; man has two eyes. . . . The macrocosm has the twelve sky signs; and man contains them too, from the head, i.e., the Ram, down to the feet, which are assimilated to the fishes." The Paracelsan doctor did not need, therefore, to study anatomy. Rather, he had to master the anatomy of the universe, the constellations, the lay of the land, and the lives of the animals. Human diseases became simply the microcosmic twins to processes observed at large: Paracelsus described a kidney stone as "tartarum because it yields an oil, a water, a tincture, a salt which inflames and burns the sick like a hellish fire, for tartarum is the hell."

The reasoning behind such a description runs something like this: Tartar is at once a name for hell and for a class of earthy substances—the crust that builds up around fermenting wine is one tartar that Paracelsus would almost certainly have known. Kidney stones are solid and earthy; they cause the torments of the damned to those who suffer from them; they produce a burning sensation—and as Paracelsus notes above, the alchemist can squeeze from kidney stones extracts that torture the suffering patient. To Paracelsus, kidney stones were not just a hellish experience; they bore a whiff of hell itself. Given a view of the body as a universe, and of disease as the extension of macrocosmic processes into the individual, Paracelsan remedies looked for ways to re-create the microcosm so as to restore the patient to order. So Paracelsus, for example, built life-size, man-shaped vessels in which to analyze urine. What happens in vessels in the shape of a man, the argument went, would replicate precisely the natural processes taking place inside a man—for Paracelsan purposes the representation of a human being became a human being, just as the human body, a representation of the universe, could become the universe, and vice versa.

The Paracelsan doctor, therefore, just like the natural magician, worked by calling into play the vital forces in the universe that could restore the harmony of what was a single, all-embracing system. Paracelsus laid bare the central assumption that made magic, like his medicine, an occult rather than a natural science: his diseases were natural processes, generated by a chain of cause and effect, but the connection between cause (the imbalance) and effect (the disease) was one of the spirit that bound together the micro- and the macrocosms. Something happened out there, which immediately evoked its echo in here. Because the microcosm and the macrocosm were inextricably intertwined, in fact, no direct physical link between those realms could or needed to be proved, for the "out there" and "in here" were already one.

The Paracelsan version of the unity of all of nature was the extreme one, the ultimate statement of what had been implied in the images of alchemy—the "rain" that a still produces, or the varied interpretations of the ouroborous, the snake biting its tail. Paracelsus's genuine insight into the connection between chem-

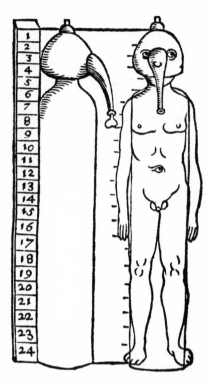

Paracelsus's "Anatomical Furnace"—a man-sized and -shaped vessel for distilling urine.

istry and disease attracted a vigorous group of students and disciples (some of whom took his reasoning to the point of using the alchemically fundamental element mercury in "drugs" that were simply poison). But the alchemical core of Paracelsan theory offered a point of attack to those who reviled both its foul-tempered author and his ideas.

Thus Daniel Sennert, writing in 1619 from a traditional medical perspective in *Chymistry Made Easie*, complained that "here we may gather that the Analogie of the great and little World is extended too large by the Chymists, because they make not an Analogie, but an identity, or the same thing. For Paracelsus requires in a true Physitian that he say this is a saphire in man, this is Quicksilver, this Cypress, this a walflower; but no Paracelsian ever shewed this." All proposed comparisons between the microcosm and the macrocosm must collapse, Sennert said, because "there is nothing so like, but in some part it is unlike." Sennert clearly identified the two great flaws in Paracelsan reasoning.

First the attempt to assert identities between the microcosm and the macrocosm, between symbols (man-shaped vessels) and the things themselves (men and women) made no sense, for the observable differences between the supposedly identical phenomena were overwhelming. Second, Paracelsan thought failed the empirical test. No Paracelsan medic ever found "a saphire in man."

Sennert's view found support even among some investigators who acknowledged their debt to Paracelsus. Jan Baptista van Helmont studied Paracelsan medicine and pursued alchemy devotedly enough to inspire the spread of rumors that he had in fact succeeded in using "a quarter grain of the philosopher's stone" to transmute eight ounces of mercury into gold. But in 1648, van Helmont flatly rejected the mapping of the universe onto the human body required by a literal interpretation of the microcosm-macrocosm image: "Away with the trifles: For we have no fountains of Salt, no reducements of venal bloud into feigned and lurking mettals. Neither are there minerals in us. . . . The name therefore of Microcosm or little world is Poetical, Heathenish, and metaphorical, but not natural, or true."

Paracelsus, in effect, found what he sought—just as those who believed in the high Greek tradition persuaded themselves that the planets moved in perfect circles. The Paracelsan concept of an alchemical universe, though, had a corrective built in: the true physician had to derive his knowledge from nature itself rather than from pure reason. And in one central area alchemists were certainly on the right track: even if their experiments failed again and again, the question they asked—what causes change in nature, why do things transform, grow, and wither?—remained a fundamental concern for the true believer and the doubter alike.

On that score, Ben Jonson could turn his sarcasm both ways, and his acid pen reproduced seventeenth-century alchemy's best defense. The alchemist Subtle asks the skeptic Surly: "Why, what have you observ'd, sir, in our art/Seems so impossible?" Surly answers: "But your whole work, no more./That you should hatch gold in a furnace, sir,/As they do eggs in Egypt." Subtle's response is precisely aimed: "Why I think that the greater miracle./No egg, but differs from a chicken, more,/Than metals in themselves." The ubiquitous examples of growth and change in nature are miracles that surprise no one. One could attack alchemical methods, alchemical reasoning—but the alchemical mission, to pursue the secrets of such miraculous processes, fed directly into the expanding ambition of the emerging natural sciences.

That ambition was realized by the occupant of the tiny laboratory perched against the wall of Trinity College. Isaac Newton, among the first of the modern scientists, was clearly one of the last of the occult masters—one who used the animating ideas of alchemy to bridge the gap between the old and new way of knowing the world. Newton's alchemical notes echo those of lesser intellects in their confusion of symbol and substance, and his own experiments seem to have been, at best, only marginally successful. At one point, in 1693, Newton managed to persuade himself that he had captured the philosopher's stone and succeeded in making gold. This "discovery" came just before he experienced what seems to have been a nervous breakdown, and following his recovery at the end of the year, Newton dropped the claim. Despite such failures, his occult studies still supplied him with an image of a universe built along alchemical lines. In a letter written in 1675 discussing the nature of light, Newton claimed that "perhaps the whole frame of Nature . . . may be nothing but various Contextures of some certaine aethereall Spirits or vapours condens'd as it were by praecipitation, much after the manner that vapours are condensed into water or exhalations into grosser substances." The next step was simple: from his spirits, Newton conceived of the idea of forces within nature that can act upon one another by means of he knew not what. By 1704, Newton would ask publicly, in *Opticks:* "Have not the small particles of Bodies certain Powers, Virtues, or Forces, by which they act at a distance, not only upon the rays of light . . . but also upon one another for producing a great part of the Phaenomena of Nature?"

In his discussion of that question, Newton, in grand alchemical style, evokes the lesson of the still, with all the fervent description of Maria's scribes. He wrote: "In Distillation the Spirit of the common Salt or Salpetre comes over much easier than it would do before, and the acid part of the Spirit of Vitriol stays behind; does not this argue that the fix'd Alcaly of the Salt attracts the acid of Vitriol more strongly than its own Spirit, and not being able to hold them both, lets go its own." In another passage, Newton reinforces the alchemical echo. "The course of Nature," he wrote, ". . . seems delighted with Transmutations." Newton retained a sense of the microcosm-macrocosm idea, supporting his claims in this section with the argument that "as Gravity makes the Sea flow round the denser and weightier Parts of the Globe of the Earth, so the Attraction may make the watry Acid flow round the denser and compacter Particles of Earth for ,composing the Particles of Salt." And finally, Newton flatly states: "It's well known that Bodies act one upon another by the Attractions of Gravity, Magnetism, and Electricity; and these Instances shew the Tenor and Course of Nature, and make it not improbable but that there may be more attractive Powers than these. For Nature is very consonant and conformable to her self."

The occult worldview permeates the whole of this discussion: look to the details; nature glories in change; the small scale emulates the large; nature repeats itself on every scale. This style of thinking is no aberration for Newton; it reappears throughout his scientific work. The law of gravitation itself rests on an assumption in which Newton's rivals recognized the taint of magic: the idea that there is a force that runs between two bodies that can act at a distance, instantly, with no direct mechanical link between them. Gottfried Leibniz heaped scorn on Newton for superstition thus elevated to the status of a law of nature: Newton's gravity, he wrote, "must be a scholastic occult quality or the effect of a miracle." Newton agreed, more or less, making one critical distinction: "These Principles I consider, not as Occult Qualities ... but as general Laws of Nature ... their truth appearing to us by Phaenomena, though their Causes be not yet discover'd. For these are manifest Qualities, and their Causes only are Occult."

In that thought, Newton bridged what we see (though he did not) as a divide between ancient and modern ideas of science. The occult survives in Newton's thought in three forms, ideas that in some shape survive to the present day. The grand concept of the unity of large and small, of the connection of every detail of experience to every other detail and to the whole, remained for Newton (as it does for us) more than a metaphor, less than a literal truth. Newton's single greatest leap of insight, probably, was his recognition that the force of gravity that produced the fall of the mythic apple was the same phenomenon that governed the motion of all astronomical bodies. His theory of gravitation was a universal, omnipresent attribute of nature—an example of the high tradition of Greek science that looked for single, simple rules that organize the entire universe. But Newton's recognition that this single principle applied to the microcosmic apple in exactly the same manner as it guided the macrocosmic dance of the planets rested on the belief, rescued from alchemy, that the ordinary affairs of this little world are directly connected to the great world of the universe as a whole. That was an assumption many Greeks and many of those who followed the classical authorities were unwilling to make. The Newtonian credo that natural laws must apply throughout the universe, and that they do so consistently (we expect gravity to operate tomorrow as it has today) is so basic to the practice of modern science that the notion is simply understood, almost never stated. That a good magician would have said the same is rarely remembered.

Next, Newton's recognition of the existence of secret, occult forces in nature freed him from a trap in which less magically inclined scientists found themselves. Whereas such men as Leibniz held out for an explanation of gravity that would include an account of what makes it work, the cause that produced gravity's observable effects, Newton was able simply to postulate an attractive force, and then ignore its inner workings. It did not matter that he could not dismantle the engine that powered gravity, so long as his account of gravity's qualities successfully accounted for the observable experience of the universe. That segregation of observable phenomena and unknown causes, and the assignment of occult status to the latter, defined the field of modern science: its task was simply to complete the description of the phenomena, the effects of the causes that transformed one state in nature to the next. Modern scientists still rely on the occult to save them from tasks they cannot master: for example, they do not ask what makes an electron both wave and particle but only how its wavelike and particlelike qualities manifest themselves.

Newton himself, though, could not quite abandon the search for underlying reasons; the scientist in him, committed to the simple description of nature, coexisted with the magus, still seeking the underlying principles that control nature. He wanted to know the answer, all the answers—and that is an ambition that survived him. As Newtonians succeeded the master himself, they abandoned his alchemical quest but took on part of the dignity of the alchemist, the magus who mastered nature. The Renaissance magician that Newton aspired to be was, in Pico's words, one to whom "it is granted to have whatever he chooses, to be whatever he wills." The magus acts; the magus is the artificer (like the great magus Daedalus, inventor and maze builder). The magus, most of all, possesses a deeper, secret insight into nature than do ordinary men; from this he derives his power and his responsibility. The magician-martyr Giordano Bruno, burned at the stake for heresy in 1600, warned that the burden of the special man was heavy indeed: his job was "a most unusual and difficult one, for you wish to lead men out of their abyss of blindness into the clear and tranquil light of those stars, which we now see scattered over the dusky blue mantle of heaven in all their beautiful variety." The blind, ordinary man would be ungrateful to the point of misery, but the magus had an obligation to look and to share what he brought into view with those who possessed the gift of sight.

Newton's work recast Bruno's challenge in terms better suited to the new, would-be scientist-magus. The description of the material world in all its complexity (as the alchemist had long sought) held out the promise of complete knowledge of and power over natural phenomena. Newton knew he had not completed the picture: "Hitherto I have explained the system of this visible world, as far as concerns the greater motions which can be easily detected," he wrote, but "there are however many other local motions which on account of the minuteness of the moving particles cannot be detected." The next step was obvious: "If any one shall have the good fortune to discover all these, I might almost say that he will have laid bare the whole nature of bodies so far as the mechanical causes of things are concerned."

The method is new: Newton asks only that his successors identify the forms of motion too small for him to detect. The aim is old: learn enough, learn all there is to know, to reveal the world in all its detail, what it is made of, how it works. The alchemists—Newton among them—had not quite achieved that perfect knowledge. To overcome their failure, Newtonian science would become an endeavor of precision measurement. Where the alchemist had sought to synthesize, combine, to imitate nature within devices like Maria's still which re-created the rise of vapor and the fall of rain to a thirsty earth, the modern magician disassembled nature, dividing, counting, searching for "local motions" and minute particles. Alchemy became chemistry, that is, and the chemists embarked on a century-long effort to identify the building blocks used to make the world. Their work tested and proved an old suggestion restated by Newton himself: that matter consisted of tiny particles called atoms, whose specific properties could account for the extraordinary range of chemical phenomena. The ultimate hope, reason's dream, was that sufficiently precise knowledge would become complete knowledge—

that the quest for the atom could yield, in the end, an accurate enough picture of nature to allow the scientist, this modern magus, to predict the course of the universe at large.

After Newton, the memory of the occult roots of Newtonian science faded, became an embarrassment. The search for a modern version of the philosopher's stone—the secret that would unlock all of nature's mysteries—continued, apparently successfully. In 1735, Alexander Pope congratulated both Newton and his age in his famous tribute: "Nature, and Nature's Laws lay hid in Night./God said, *Let Newton be!* and all was *Light.*" So it was then; so it seemed to remain. The next step was the unequivocal triumph of Newton's methods; the step after that led down a path unforeseen, undreamed of in the glow of Newton's divine light.

Chapter 5

OUR POWERS

INCREASE

WITHOUT

LIMITS

Isaac Newton ended his career as an alchemist around 1696. In that year he abandoned Cambridge and the academic life, moving to London to take up duties as the warden of the mint. As warden, and later master of the mint, Newton was a scrupulous civil servant. He had attempted to manufacture gold by art, but as the king's officer, on the trail of those who tried the same trick by less subtle means, he was ruthless, personally interviewing prisoners and informers and when he could, consigning counterfeiters to prison and the gallows. One of the most notorious coiners, William Chaloner, had tweaked Newton in public for years, even accusing him publicly of tolerating fraud and abuse at the mint. Newton responded by building a network of agents and informers who tracked down Chaloner's counterfeiting enterprise and whose testimony finally convicted him of treason.

At the last, Chaloner recognized his fatal mistake. He wrote to the warden: "O my offending you has brought this upon me O for Gods sake if not for mine Keep from being murdered O dear Sir no body can save me but you O god my God I shall be murderd unless you save me O I hope God will move your heart with mercy and pitty." Newton was unmoved. On March 22, 1699, Chaloner was taken to the place of execution and hung by the neck until dead.

But while Newton pursued his campaign against counterfeiters for more than thirty years, the question he left behind in Cambridge remained unanswered. His retreat from alchemy is probably the best marker of the moment when science finally subsumed the occult as the source of what was understood to be a true understanding of the world. Alchemy at root had been an inquiry into the nature of physical reality. When it failed, the mystery remained: what is the world made out of—what is the fundamental form of all the stuff, the matter, of which the world is composed? In 1717, twenty years after leaving his laboratory, Newton was prepared to ask that question in public, as a challenge not to magic but to science. The press of public duties had already made its claim on his attention, so he did not come to any firm conclusion about the actual composition of matter. Instead, as he wrote: "Since I had not finish'd this part of my Design, I shall conclude with proposing only some Queries, in order to a farther search to be made by others."

That search reached its goal a century later, with the publication of a Swedish journal, *Afhandlingar i Fysik, Kemi, och Mineralogi* (*Transactions of Physics, Chemistry, and Mineralogy*). The fifth volume of the journal, published in Stockholm in 1818, was devoted to the work of a single author, Jöns Jakob Berzelius, who in nearly five hundred pages of results outlined one of the founding theories of modern chemistry. Many of his ideas did not survive the detailed investigations of the next two generations of scientists. But there was, almost lost amongst the longer papers, one slim, seven-page report that still inspires awe among laboratory researchers. It has become one of the few classics of scientific literature, for in it Berzelius settled (for a time) the question of the inner architecture of matter.

Throughout the history of the investigation of nature, there have been two main answers to that question. One says that the world is all of a piece, formed from the same stuff, a single kind of matter that can be rearranged to build all the variety that we see. The other holds that the world is made out of many different stuffs, matter that in its simplest state retains the distinct qualities that produce all the varied substances we encounter in the ordinary world of experience. Berzelius's work supplied the clearest confirmation to that date that the second answer was the right one, that matter exists in nature in a dizzying array of fundamental forms. Specifically, Berzelius reported on the results of over a decade of precision measurements with which he identified the relative weights of the atoms of forty-five of the forty-nine elements then known. Berzelius was famous as Europe's premier virtuoso of experimental technique, and many of his final values for atomic weights are strikingly close to the best current figures: his lead weighed in at 207.4 (relative to the figure for oxygen, set at 16), compared with the current figure of 207.2; he fixed chlorine at 35.47, compared to today's 35.46; nitrogen was measured at 14.18, and is now recorded at 14.01—and so on. The discovery of clearly identifiable atomic weights for each element finally confirmed that the chemical atom really existed: that every element was made out of discrete chunks of matter, each as distinct from the next at the smallest level—that of the atom—as an ingot of silver differs from a lump of lead.

The idea of the atom completed the chemists' escape from their alchemical roots. The alchemists' still had mimicked nature in order to alter it, to change lead into silver, silver into gold. That would be impossible if an atom of gold were irreconcilably distinct from one of lead—exactly what the new view suggested. Freed from the fruitless task of transforming atoms, then, chemists could begin to account for them, and the archetypal tool of their trade was the balance beam, the precision scale that could weigh the world into its component parts. Newton had asked his successors to identify the behavior of matter on the smallest level. With Berzelius's description of the chemical atom, they seemed poised to do so. Berzelius himself turned his scale into a scalpel that could slice apart a compound with a precision no alchemist ever dreamed of, distinguishing one atom from the next with measurements accurate into the range of one part per 10,000. It was, Joseph Priestley boasted, following his discovery of "pure" air (oxygen), as if "our powers of investigation . . . seem to increase without limits."

But with the identification of the chemical atom, Berzelius and his colleagues achieved a kind of high-water mark in the drive to know without limits on the knowing—a last clear victory of the scientific revolution as Newton had shaped it. The methods of both measurement and thought that the pioneer chemists developed did in fact trap the individual properties of matter, atom by atom. In doing so, however, those chemists undermined the belief that had governed the search for the true theory of matter since it began among the Greeks: that the human mind could create a comprehensive description of nature as a whole out of the accumulation of unassailably true individual facts. In the early years of the nineteenth century, the pursuit of the atomic hypothesis successfully demonstrated that nature is knowable piece by piece. But contrary to Newton's hope, the persistent increase in the precision of measurement could not build from those pieces the complete understanding of nature that was the ultimate goal of his science. The identification of the chemical atom begins, that is, as a story of the triumph of the Newtonian belief in the perfectability of human knowledge. But in the end the tale turns: instead of proving the unlimited power of reason, the atom's story becomes that of the modern attempt to understand where the limits to our knowledge lie.

That story starts, like much in science, with the strangely inspired speculation of a group of Greek thinkers: Leucippus and his student Democritus, who flourished in the fifth century B.C., and Epicurus, working in the fourth century B.C. It was recorded by a Roman, Lucretius, who described the theory in astoundingly elegant metrical verse in his epic poem *On the Nature of Things* (thus setting a literary standard for scientific writing few have met since). Their concept of the atom emerged out of an act of faith: without any hope of proving it, the atomists believed in what we now would call a conservation principle, that of the conservation of matter. As Democritus put it, "Nothing can be created out of nothing, nor can it be destroyed and returned to nothing." (Substitute the Einsteinian concept of mass-energy for matter, and the principle remains one of the core axioms of physics today.) The original atomists came to their belief in the conservation of matter by thinking about what might happen if one attempted to slice any visible piece of matter in half, then in half again, and again, indefinitely. Greek geometers had already established the mathematical idea that a line could be endlessly divided into an infinite number of segments: What about the real world? Faced with a physical object rather than an imagined one, the atomists concluded that endless division would produce cosmic mush, formless and ultimately nonexistent. Most important, such formlessness would be (the atomists argued) incapable of reconstituting itself into the entire bestiary of different shapes and properties displayed by all the real stuff we experience every day. But if matter could neither be destroyed nor divided forever, then at some point the dissecting knife would have to stop, its edge blunted against some indigestible chunk of matter. The point at which matter could not be divided further was, the atomists concluded, the stuff of nature in its simplest form, particles too small to be seen—those invisible irreducible specks they called atoms, from the Greek *atomos* (indivisible).

To explain how these invisible atoms, all made of the same stuff, differing from each other only in shape and size, could build the world, the atomists created one of the most beautiful metaphors in the history of science: their atoms accumulate into nature as the letters of the alphabet can be transformed into poetry. Atoms by themselves, like letters, are isolated, without meaning. Atoms in combination mimic words, utterances, becoming compounds with distinct qualities, which Epicurus named molecules. Molecules assemble to produce fully developed macroscopic phenomena—people, things: a sentence, nature's thoughts realized. Differences in shape or texture, the change in the color of the sea as the sun sets, could be accounted for by the rearrangement of the "letters" to produce new "words" and a new text. Lucretius described this natural poetry in a poem of his own: "Obviously it makes a great difference in these verses of mine in what context and order the letters are arranged. If they are not all alike, yet, most are so; but differences in their position marks the difference in what results [the words]. So it is when we turn to real things; when the combination, motion, order, position, shapes of matter [the atoms] change, so does the thing composed [out of those atoms]."

With this hierarchical idea of the organization of matter in nature, the atomists achieved one of their critical goals—bridging the gap that had already appeared between one-stuff and many-stuff theories of the fundamental composition of the universe. One of the competing ideas, proposed by Thales of Miletus among others, held that all of nature was simply made of one continuous substance—Thales suggested water—whose eddies and burbles could produce the varied world human senses experience. Against this idea of constancy, rival thinkers noted that the constant ebb and flow of such a cosmic ocean could not produce what seemed to them to be the fundamental differences seen in the diversity of a cosmos that stretched from rocks to beasts, from beasts to stars. Such thinkers proposed instead that all matter was composed out of four cardinal substances: earth, air, fire, and water. Those four elements could then blend in an infinite number of combinations to create all the different phenomena that could be observed.

Until the idea of the atom emerged, that is, one hand paddled through water, water everywhere, while the other clutched a fistful of earth, a touch of fire, water, air, all jumbled together in an unrelenting tangle: two irreconcilable portraits of reality. With the atom, however, the two conceptions could merge. Atomists preserved the vision of an underlying unity of all matter—all atoms, after all, were made of the same fundamental stuff. At the same time the atoms did differ from each other in shape, size, and motion, just as letters differ from one another, which allowed the atomists to create a detailed explanation for the existence of the specific properties of different substances. For example, Democritus suggested that a metal like lead possessed the quality of heaviness because it contained either more or larger atoms per volume than lighter materials like iron. With the same kind of reasoning, Democritus accounted for iron's strength, relative to soft and malleable lead, by suggesting what differed was the organization of each metal's atoms in space. Atoms of lead, he suggested, could arrange themselves in regular lattice. By contrast, iron atoms could clump irregularly. The greater density of small bunches of iron atoms scattered through each lump of the metal, Democritus argued, is what makes iron hard, as compared to lead.

It was a fine guess: inspired, for with his notion that the smallest scale organization of matter shapes the attributes of the whole, Democritus glimpsed the concept of microstructure that we still use to analyze materials; a guess, for Democritus and his fellow atomists lacked both the data and the interest in making the observations they needed to prove their ideas. Ultimately, of course, Democritus's atoms bear little resemblance to the atoms that the nineteenth-century chemists identified, and they differ again from the atoms we now perceive. That is no surprise, of course; science moves on. But the difference serves as a reminder: the Greek atomists were philosophers, not scientists, and the point for them was not to find a fact but to seek a larger truth. Democritus was a pessimist. To him, human senses were fallible, deceiving, and ultimate knowledge lay beyond them. "We know nothing in reality," he wrote in one tantalizing fragment, "for truth lies in an abyss."

What Democritus sought in that abyss was not an actual physical theory with which to dissect the world but an idea that would allow him to imagine it. Lucretius, who followed Democritus's successor, Epicurus, did believe that atoms were real, but his aim was essentially the same: his atomic reasoning did not lead him to any laws of nature. Rather, he was after some clues about the meaning of life. His atoms conferred a kind of immortality, he wrote, "for that which once came from earth to earth returns back again. . . . Nor does death so destroy as to annihilate the bodies of matter, but it disperses their combination abroad, and then conjoins others with others." But though atoms traipse through an eternal dance, coming together and falling apart, the human soul did not. Lucretius wrote: "Even if time shall gather together our matter after death and bring it back again as it is now placed, and if once more the light of life shall be given to us, yet it would not matter to us that even this had been done, when the recollection of ourselves has once been broken asunder, and to us now, no memory, no anguish remains from those who we were before." Therefore, Lucretius concluded: "We may be sure that there is nothing to be feared after death"—for both sensation and memory are lost forever. Life is lived once, now, and that is all.

It was a simple credo, albeit austere. Greek atomism was founded on a leap of faith (matter is conserved, eternal and inviolate); it culminated here, in faith again (act now, for only matter is eternal—not man nor men). Belief in such an idea depended on how persuasive such faith seemed—each of the Greek theories of matter convinced those who responded to them by appearing too elegant to be false. Distinguishing between rival notions was not a science but an art, an aesthetic choice. As a matter of practice, the four-element theory of Empedocles dominated late classical and medieval concepts of matter. The alchemists in particular seized the concept of mixtures they could manipulate and transform. Lucretius's poem preserved the memory of the atom, but little more—and in Christian Europe his doctrine of the mortality of the soul was rank heresy. Those first atoms were almost ghosts, creatures entirely imaginary, an image that convinced a few hardheaded philosophers and no one else.

But within the original concept of the atom lay the seeds (to use a word Lucretius often equated with his atoms) of a renewed atomic theory. The first was simply the idea itself: that it was possible to conceive of an indivisible, simple particle that could exist beyond the reach of human senses. The second thought was perhaps more important, for pondering the connection between atoms and everyday life forced the first atomists to confront the central problem facing anyone seeking to understand the nature of matter: How does one perceive, study, what human senses cannot detect—how could a science of matter render intelligible what remains invisible? Early matter theories exploited simple analogies: water in a riverbed remains water, yet takes on different shapes—therefore the universe is made of water, or something like it. By contrast, in Lucretius's account, the atomists recognized both the possibility and the need to reason, step by step, from what they could see to what they could not—to build a chain of inference that could bridge the gap between experience and the hidden essence of matter. Thus Lucretius again and again offered examples of macroscopic phenomena that could lead toward an idea of the atom, most famously in his explanation of what resembles what we now call Brownian motion:

> It is proper to give attention to these motes that are seen tumbling in sunbeams, for their motion is an imitation of underlying movements hidden from sight. There you will see many particles set in motion by invisible blows, changing their course and beaten back, this way and that, in all directions . . . [this motion] originates with atoms, which move of themselves. Small compound bodies nearest the impetus of the atoms are set in motion, driven by the invisible blows of the atoms, while they in turn attack those that are a little larger. Thus the movement ascends from the beginnings and by successive degrees emerges upon our senses, so that those bodies which are moved we can see in the sun's light, moved by blows that we cannot see.

Lucretius was wrong—dust in sunbeams dances to the tune the wind blows, not to the microscopic jitter of the atoms—but he was almost right, and the coincidence is striking. In the nineteenth century, with the aid of a microscope, true Brownian motion was discovered in the random paths traced by particles suspended in liquid. Its formal, mathematical description came in 1905, when Albert Einstein solved the statistical problems raised by the phenomenon. Certainly, though, Lucretius himself was no scientist as we recognize the breed, and his argument was intended to persuade, rather than to demonstrate. But he was not simply blindly guessing, and his reasoning contained the suggestion that the hidden world could be reached, not directly, but by stealth and skill, with a kind of judo of the mind. Lucretius himself did not trace out the implications of his form of reasoning, this careful, incremental weave of observation and interpretation that could tie imperceptible phenomena to their observable consequences. But the hints he left behind offered a glimpse of the ultimate potential of atomist ideas. If atoms were real, a combination of human reason and careful observation could chase them down. If they could be found, it followed, the world could be described, known in every detail, down to its fundamental constituents, the building blocks of matter.

Lucretius died around A.D. 55. The next gasp of interest in the atomist idea in Europe did not come until the sixteenth century, when the corrosive ideas of Copernican astronomy, along with a wave of other, novel investigations, prompted the reexamination of old ideas. Giordano Bruno was the first Renaissance figure to pick up the notion of atoms while pursuing his own, peculiar brand of magic. When the Inquisition finally burned him at the stake in 1600, his heresies included the claim that between his "minima" (atoms) lay emptiness, void. In the seventeenth century, Galileo wrote about them, as did Francis Bacon and Thomas Hobbes—but atoms remained in their hands essentially as they were when Leucippus first imagined them: a mental picture that might be true, rather than a physical fact. That was how the next great atomist conceived of them, too, but who he was conferred on what he said an authority none of the others could match. Isaac Newton proposed his atomic hypothesis, stripped of any of his secret, alchemical convictions, in a form that closely resembled the original Greek idea: "God in the Beginning form'd Matter in solid, massy, hard impenetrable, moveable Particles, of such Sizes and Figures, and with such other Properites . . . as most conduced to the end for which he form'd them." Like his predecessors, Newton was guessing—as he admitted, sort of: "At least," he wrote, "I see nothing of Contradiction in all this."

But when Newton mused, it was with a purpose. For the Greeks, atoms were the foundation of a worldview—an answer to the question What is the world made of? Newton took the same idea and asked a different question: Given the atom, what then? Assuming the atom to be real, Newton posed the problem of finding out what form it takes, how it behaves, and what laws it must obey to produce the sensible world he had already analyzed with his laws of motion and gravitation.

What Newton realized was that to find the atom, one first had to look for it, imagining what one hoped to discover. Newton himself used the assumption that atoms exist to define the relationship between the pressure and density of a gas made up of individual, mutually repulsive particles—thus producing the first quantitative law of the atomic age. With the sanction of the master, the contemplation of nature in its component parts ceased to require a leap of faith and became instead a question of analysis, a task of measurement. The hunt for the atom began in earnest during the century following Newton's death. Its ultimate capture occurred as a generation of chemists finally learned how to pry apart the world with a scale.

The tool that would trap the atom was and is one of the simplest instruments still employed by modern science: the two-pan balance—those same scales that blind justice holds. The balance is almost as old as human culture itself, and its basic principle has remained unchanged since it emerged at least three thousand years ago. A horizontal beam is suspended on a pivot fixed precisely in the middle of the beam. Identical pans hang from each end of the beam. Unloaded, or loaded with equal weights in the pans, the beam remains level. If the weights in the two pans are unequal, the beam tips, and the angle it reaches indicates the difference in the weights of the two loads. The oldest scale discovered includes each of these features: an Egyptian balance dating from 1400 B.C. consisted of a beam hung from a hook with a cord, with loops of string at either end of the beam to accommodate the pans.

Such scales were not terribly sensitive to small differences in weight, but such crudity did not matter much so long as the main concern was honesty, not precise accuracy. The history of the

A drawing of Egyptian scales from the books of the dead.

balance beam is a history of commercial fraud, which, like the instrument itself, dates back as far as human memory extends. One simple method of cheating is to pivot the beam on a point slightly off center.

The more difficult fraud to catch is (not surprisingly) the more common, at least as the history of prohibitions against it would suggest. Laws against false weights—one labeled a pound that weighs fifteen ounces, or seventeen, depending on the terms of the sale—appear in every ancient commercial code. The Bible warns in Deuteronomy: "Thou shalt not have in thy bag divers weights, a great and a small." And, driving the point home: "Thou shalt have a perfect and just weight." If further clarification were needed, Proverbs adds the admonition that "a false balance is abomination to the Lord." Sharp dealing aside, the drive to increase precision probably emerged from the need felt by traders in gold and precious stones to measure small differences in weight accurately. The Arabs used the seed of the coral tree, called a *qirat*, to measure the weight of jewels, yielding the unit now called the carat, standardized at two tenths of a gram. Alchemists also pushed the development of sensitive balances, especially with the Paracelsan turn toward medical alchemy, based on the precise construction of particular compounds to make drugs tailored to each disease.

The improvement of the scale to jeweler's standards involved no advance in theory, in the principle of the instrument. Rather, it was a problem of fine workmanship. The beam had to be precisely machined so that it had a constant weight and strength throughout its length, and the same was true for the pans and the system that connected them to the beam. The pivot on which the scale balanced had to be placed exactly, and designed so that it turned smoothly and evenly in either direction. All of this was well within the capacity of seventeenth- and eighteenth-century instrument builders. But before they invested the time and effort required to make still more sophisticated devices than those required by gold dealers and physicians, though, some new clientele had to emerge willing to pay top prices for top quality.

The Newtonian injunction to measure the forces of nature precisely created that demand—or rather strengthened the appetite for precise measurement that the emerging community of scientific investigators already felt. Galileo, making his own telescopes and Leeuwenhoek, manufacturing microscopes by the dozen, reflected the state of affairs early in the scientific revolution: if one needed an instrument built right, best build it your-self. But the scientific revolution was genuinely an upheaval, not just in thought but in social practice as well. The increasingly large number of clearly professional scientists prompted the emergence of a whole raft of new professions. The instrument-building craft left the artisan's workshop and became an industry in service of the new science, itself practiced on an industrial scale. Jesse Ramsden, one of the best of the British builders in the eighteenth century, could not keep up with the demand for his instruments with a staff of sixty. Ramsden's work was expert enough to earn him the status of a man of science in his own right, gaining him election to the Royal Society in 1786. (He was also the butt of a famous story: ordered to appear before King George III, he arrived at the palace at the correct hour and day—one year late.)

Ramsden's precision balances achieved an accuracy of at least one one-hundredth of a grain—about .000023 of an ounce. That was good enough to change chemistry forever. Antoine Lavoisier, tax collector and chemist, allowed as how there were but two types of scale worth owning for serious work: those built by a Parisian maker, Nicolas Fortin, and those produced by Jesse Ramsden. Lavoisier, born in 1743 and executed for his role in the *ancien régime*'s taxing apparatus during the Terror in Paris in 1794, used his balance to achieve the final liberation of chemistry from its alchemical roots. His investigations examined and discarded each of the ancient four elements in turn, but his study of com-bustion, of the chemistry of fire, yielded both a new method of analyzing the world and specific discoveries that shaped the search for the atom.

Perhaps the single most self-evident fact about fire was that when something burns, the flames escape from the substance, and the combustible material loses some of its bulk. But what is that missing stuff? By the time Lavoisier confronted the prob-lem, the classical image of fire as an element—a constituent of matter—had already been revised. Eighteenth-century chemists thought of fire and its effects on matter as a cross between an active agent and a kind of element, a component of compounds. To explain combustion, they invented the concept of phlogiston, an insensible, undetectable fluid. Anything that could burn held phlogiston and could absorb it from its surroundings. When heated or set alight, such substances would give off or transfer phlogiston. All chemical transformations taking place in a fire—

the conversion of metallic ore to pure metal in the smelting process, for example—resulted from the movement of phlogiston, an exchange in this instance between the charcoal and the ore.

A genuine virtue of the idea, as the Nobel laureate chemist Roald Hoffmann has pointed out, was that it provided a holistic view of chemistry and chemical transformations, focusing on an entire system—fire, charcoal and metal, in this instance. With that focus, chemists could maintain a kind of phlogiston budget for such systems—an accounting that explained chemical changes as an ebb or flow of this elusive fire principle. But by the second half of the eighteenth century, an accumulation of observations were beginning to vex such calculations. Paracelsus and his followers argued for a link between fire and air throughout the sixteenth and seventeenth centuries—and several of their experiments proved that fire could not take place in a vacuum. But most people simply ignored the hint contained in such experiments until an episode of scientific vaudeville took place in Paris in 1768.

The show opened with a startling piece of news: diamonds— thought to be the most indestructible of objects—could be destroyed by fire. The vulnerability of diamonds to heat had actually been observed at least as early as 1695, when two men in the employ of Grand Duke Cosimo III of Tuscany used a burning lens to heat a diamond. But despite the existence of a few scientific journals, most seventeenth- and eighteenth-century scientific communication took place in letters from one scientist to the next—and the diamond experiments went unreported. (After the fact, it turned out that the Tuscan experiment had been repeated several times between 1695 and 1768, but an observation becomes a discovery only when someone besides its discoverer notices it.) When the French Academy of Sciences at last heard that diamonds could be consumed, it came as a shock. By 1770, diamonds had been destroyed on a public stage in Paris, leaving no room for doubt that the phenomenon was real. But that left an undodgable question: What happened to those vanishing diamonds? In 1772, the academy put together a panel to find out. Lavoisier, aged twenty-nine, was its youngest member.

The team first tested the idea that diamonds are somehow volatile—that they could evaporate like the ice that becomes steam when heated. To test this hypothesis, Lavoisier hit upon the idea of trying to distill the spirit of diamonds. He and his colleagues took diamond chips and small whole stones and heated them in stoneware vessel for three hours. At the end of the run, the diamonds had lost about 14 percent of their weight, but the still's receiver contained nothing but a little bit of water. With that issue apparently settled, the team moved on to the next experiment. The first tales of the destruction of diamonds claimed that heat obliterated diamonds in both open and airtight vessels. But a local jeweler named Maillard believed otherwise, holding that without contact with the air, diamonds were safe, and he put his money behind his hunch. Maillard donated three stones to the academy scientists, on condition that he could prepare them for the furnace. He placed the stones in clay tubes, packed them tight with charcoal granules and sealed the ends. The tubes were then placed in a crucible, packed with powdered chalk, and that crucible was fitted within two larger vessels, sealed together. This elaborate package went into a furnace, heated to the highest temperatures attainable, and then unpacked. The packing itself deformed with the intense heat, but the diamonds within were intact.

Lavoisier wrote the team's report to the academy, concluding that diamonds were involatile. The second experiment led the team to state that the destruction of diamonds appeared to be a form of combustion, "just like that of carbon"—a finding the group promised to test further. Lavoisier himself apparently disagreed with some of his colleagues, believing still that they had not ruled out the possibility that one could evaporate diamonds. But the phenomenon itself contained a hint: fire and air clearly had some bond of chemical action.

Alchemists had long experimented with roasting metals over high heat—lead and iron, copper, tin and mercury. They found that they could produce a fine powdery substance, often different in color from the original metal, which they called a calx. Following traditional reasoning, such calces formed as a result of the expenditure of phlogiston, given off by the metals as they heated. But a troubling anomaly kept turning up, one finally confirmed, as far as the academy was concerned, in early 1772. In a series of carefully controlled experiments Louis-Bernard Guyton de Morveau showed that the calces of the metals weighed more, not less, than the original samples. If those calces had lost phlogiston as they formed, then this insensible fluid would have to have a negative weight, some quality like the Aristotelian idea of "lightness."

The implication, which Lavoisier recognized almost immediately, seems obvious after the fact: rather than load the mysterious phlogiston down with an ever more elaborate catalogue of unique properties, look for a simpler solution. Beginning in 1772, Lavoisier carried out a series of experiments that destroyed once and for all the phlogiston theory replacing it with one that finally recognized the true relationship between fire and air. In this final set of tests, Lavoisier made one crucial assumption: he accepted as a fundamental principle the old atomist idea that matter is conserved—and hence that the weight of any system under study would remain constant, no matter how many times chemical reactions changed its form. Given that assumption, the test of the mechanism of combustion became a simple task of precision measurement.

The experimental series began with the burning of phosphorus and sulfur, two easily combustible elements with which Lavoisier showed that the the flaming samples absorbed air from the surrounding atmosphere. The next major experiment led Lavoisier to his ultimate breakthrough. To believers in the phlogiston idea, air was a simple substance, one stuff. A substance that Lavoisier would later name oxygen had already been discovered—but, under the label "dephlogistated air," it had already been wedged into the older system. While this dephlogistated air caused fires to burn more brightly, its relationship to atmospheric air was explained by suggesting that the oxygen was simply purer air, containing less phlogiston (and hence capable of absorbing more from a burning fire), than normal air. Lavoisier set out to prove that air was actually a mixture of two different substances, one that was essential for combustion, and one that played no role in fire.

To do so, Lavoisier performed the experiment that became a model of the method that could separate chemical elements from compound substances. Following the same tack that the English chemist Joseph Priestley had originally used to isolate oxygen, Lavoisier placed four ounces of mercury inside a jar that enclosed precisely fifty cubic inches of ordinary atmospheric air. He heated the mercury for twelve days at a temperature just below the boil, driving the reaction that produced a calx of mercury, its crust of characteristically red flakes. At the end of the twelve days, Lavoisier extinguished the fire and allowed his apparatus to cool. The volume of the air, he reported, was down about one sixth, and the air left in the vessel extinguished candle flames. Next, using his prized precision balance, he found that his calx of mercury together with the mercury itself weighed forty-five grains (approximately .09 ounce) more than the original pure metal. Finally, Lavoisier reversed the process. He collected the forty-five grains of the red calx in a small glass vessel with a neck that he inverted in a jar filled with water, thus preventing the intrusion of air. He then heated the calx quickly, past the boiling point, which reduced the calx back to its pure form, mercury. A small quanitity of a gas bubbled up through the water, almost the same amount as had been lost in the first go round. When that gas was added to the diminished, candle-killing gas first produced, Lavoisier wrote, "we reproduce an air precisely similar to that of the atmosphere, and possessing nearly the same power of supporting combustion and respiration, and of contributing to the calcination of metals."

That is, Lavoisier showed that the processes of both combustion and calcination were reactions that involved the fixing of one gas (oxygen) out of the mixture of two gases (oxygen and nitrogen) that make up common air. He did so by showing that though the weights of the gas and the solid portions of the system varied, the total weight remained the same throughout the entire sequence. Phlogiston had no role to play and could be abandoned. Lavoisier then broadened his attack on the phlogiston theory, proving that the combustion of charcoal and pure oxygen produced another compound gas, "fixed air," or carbon dioxide, and that water, too, was a compound of two gases, oxygen and "inflammable air"—hydrogen. To make each of these discoveries, Lavoisier heated controlled quantities of the compounds under study and weighed the changes in the amount of gas and the amount of solid or liquid at each stage of the transformations that followed. Given his assumption of the conservation of matter, confirmed each time the total weight in one of his experiments remained constant, the step-by-step measurements allowed Lavoisier to construct a clear picture of elements that combine, separate, and recombine in precise proportions.

With the final confirmation of the oxygen hypothesis, the classical theory of the four elements, already battered, suffered a killing blow. Fire could no longer be thought of as a constituent of matter; rather, it was the signal that a chemical reaction was taking place. Air lost its elemental character too, transformed by analysis into a welter of "airs," gases with distinct properties. Water decomposed into two such gases, and, as for the earth—

the earth abides, but in Lavoisier's new chemistry, the distinctive qualities of each element defined them far more precisely than any presumed common earthiness. Put another way: in about a decade, Lavoisier completed the destruction of a description of nature that had stood for almost two thousand years.

Lavoisier announced his findings to an international public in 1789, with the publication of his *Traité Élémentaire de Chimie*. The English translation followed in 1790, along with editions in all the major European languages; it became for chemistry what Newton's *Principia* was to physics: both a textbook and a manifesto. At first glance the book simply reports the outcome of a number of experiments, laying out the series of steps by which Lavoisier reached his understanding of oxygen and the other gases his study of combustion revealed. To practicing chemists, this account of scientific results mattered a great deal, for it provided a body of work to test and extend. But the book was from the first far more than a research report. Just like Newton, Lavoisier sought to use the revolutionary nature of his results to construct a whole new approach to science, to be defined in the simplest terms: "Chemistry, in subjecting to experiments the various bodies in nature, aims at decomposing them so as to be able to examine separately the different substances which enter into their composition."

One of Lavoisier's major themes was that this new chemistry had to be strictly empirical. Occult notions, like that of a mysteriously light, undetectable phlogiston, had no place in science, he argued, in language that almost precisely echoes Newton's. "It is a maxim universally admitted in geometry," Lavoisier wrote, "and indeed in every branch of knowledge, that, in the progress of investigation, we should proceed from known facts to what is unknown." What counted was the record of the precise quantities obtained from beginning to end of an experiment. At bottom (for that part of chemistry that has driven away the less meticulous ever since) the pursuit of "what is unknown" was simply the pursuit of detailed, accurate measurements, no more and certainly no less. In this light, Lavoisier saw his instruments, his scale, as more than mere laboratory tools; the precision balance was the ultimate source of chemical knowledge. He proclaimed: "The best method hitherto known for determining the quantities of substances submitted to chemical experiment, or resulting from them is by means of an accurately constructed beam and scales, with properly regulated weights, which well known operation is called weighing . . . the usefulness and accuracy of chemistry depends entirely upon the determination of the weights of the ingredients and products both before and after experiments."

Chemical secrets lurked in the details—Lavoisier had found them there, and others, he urged, could do so too. More accurate measurements were not simply better than grosser estimates; the highest standard of accuracy was a requirement, not a luxury, for only with precise measurements could the behavior of "the various bodies of nature" be tracked with confidence. The occult disciplines, alchemy, had been called arts as well as sciences: the individual (and in some sense unique) genius of the artist was what produced the desired effect. In rejecting the older traditions Lavoisier built a chemical method that was a craft, with techniques to be mastered, available to any practitioner. In his treatise, consciously written as an elementary text, Lavoisier laid out his vision of a science in which discoveries emerged from a method of inquiry accessible to all those disciplined enough to weigh with care.

But Lavoisier's doctrine of accurate measurement was only half of his conception of the new method that could reveal fundamental truths of nature. The facts provide a start, but to understand what the facts mean one had to learn anew how to process them; how to arrange and how to interpret them. Language mattered: what one called chemical compounds, reactions, processes determined, Lavoisier argued, how one thought about them. He cited a logician, the Abbé de Condillac, who wrote that "the art of reasoning is nothing more than a language well arranged," and argued further on his own that "as ideas are preserved and communicated by means of words, it necessarily follows that we cannot improve the language of any science without at the same time improving the science itself; neither can we, on the other hand, improve a science, without improving the language or nomenclature which belongs to it." To that end Lavoisier coined terms as prolifically as he could—oxygen itself being the most famous—replacing a welter of balky phrases like "dephlogistated air" or "eminently respirable air."

More significantly, Lavoisier invented the fundamental form of chemical argument, the chemical equation. Lavoisier had inherited a language—and hence a science—that emphasized qualities: "dephlogistated air" is a term that emphasizes what that substance does, what it is (or seems to be). The language, and hence the science, that Lavoisier created centered on quantities, numbers abstracted from the particular properties being studied. His axiom, that matter is conserved, led him to express each reaction, each chemical event as an equality, a relationship between amounts of stuff. The state of affairs "before" lay on one side of the formula, with the outcome "after" on the other. The first chemical equation ever written described fermentation, a reaction near to Lavoisier's heart: some quantity of grape juice, a number = quantities of carbonic acid + alcohol, numbers both.

What emerged directly from this new chemistry, composed equally of new techniques and revolutionary language, was, at last, the creation of a coherent description of matter: the identification of the atom itself. As it happened, Lavoisier himself did not accept the atomic hypothesis. He argued instead for the idea of multiple elements, which he defined as the simplest forms of matter accessible to experiment, saying nothing about whether such elements broke up into discrete chunks or formed a continuum. But his new language—or perhaps better, his new syntax, this quantitative structure of chemical thinking—could finally frame the right kind of question, one that could yield genuine insight. Newton himself in his version of atomism understood that the original Greek query (Do atoms exist?) led to a dead end. Lavoisier's innovations enabled his successors to accomplish what Newton had hinted they should attempt, to presume the atom, and to ask: "If atoms exist, how would they behave?"—and then further, to construct the experiments that could yield an answer. Revolutions in science are made not just from surprising results but from the construction of whole new ways of thinking, of asking the questions that can lead to discovery.

The revolution in the science of matter was driven from Paris in 1794. The political revolution swept it away, claiming Lavoisier on May 8. His colleagues made some effort (not much—it was too dangerous) to save his life, but his ties to the *ancien régime* condemned him, and the president of the court that committed him to the guillotine in the place de la Révolution dismissed the pleas on his behalf with the famous (and possibly apocryphal) aphorism: "La République n'a pas besoin des savants" ("The Republic has no need of experts"). But though the French Revolution ultimately failed in its attempt to export itself, Lavoisier's scientific insurrection had transcended national boundaries long before the Terror caught up with him. The discovery of the atom proceeded in two stages: first, one man invented a theory that required the atom; then another man performed the experiments that proved the theory. The latter investigator was the most famous professional chemist of his day, the Swedish researcher Berzelius; the former was an obscure, self-taught Englishman named John Dalton—the first man to frame the problem of matter in a form in which it could be solved.

Lavoisier had been brought up in comfort. He was educated in Paris, receiving both formal classical training and detailed instruction in the sciences. As a young man he had money enough and time to pursue his own interests, and moved naturally into the elite circles of scientists, earning election to the French Academy of Science when he was twenty-five. Dalton, by contrast, was born in a Cumberland village in 1766. His father was an artisan, a hand-loom weaver, whose occupation was about to be eliminated by the spread of water- and then steam-powered textile mills. Dalton's family were Quakers, and the young John Dalton studied at a Quaker school in the village. That education served him well enough: while many more orthodox schools continued to emphasize Latin and Greek, Dalton's dissenting schoolmaster taught him mathematics. Having completed a course on "surveying, mensuration, and navigation," Dalton left school—aged eleven.

Dalton's talent manifested itself early, but he was a poor boy who had to earn his keep. At twelve he opened his own school, and at fifteen he joined the faculty at his cousin's academy, continuing to study informally with older masters as necessary. By the time he turned twenty-one, Dalton had made his own way in the world for a decade, as generations of poor children had done. What made him extraordinary, at least to modern eyes, was that while working in almost complete isolation, spending most of his childhood simply making a living, Dalton managed to turn himself into an active, practicing research scientist. That achievement did not seem as improbable to him and his contemporaries as it does now. One of Dalton's mentors at his second school taught him more mathematics and introduced him to the field of meteorology; at his urging, the young Dalton simply began to study atmospheric phenomena, as an ordinary, perfectly appropriate avocation. His pursuit of what was then fundamental science portrays in a kind of intellectual microcosm a sense of what animated both the international society of science and his own, rapidly transforming Britain: nature had never been more intelligible, more accessible to human reason, a book open even to a half-trained twenty-something children's teacher from the hinterland.

Dalton's choice of what to study produced one of the peculiar coincidences that dot the history of science: both Lavoisier and Dalton began their scientific careers by studying the weather. Both men, from very early in their lives, gathered meteorological data, collecting measurements of barometric pressure, temperature, and the like. Lavoisier did not pursue the subject very deeply—he wrote one paper on forecasting, and let it go at that. Dalton persisted, however; off in the countryside he had weather all around him, and little else. In this earliest phase of his work, he attempted to account for the circulation of the trade winds, to correlate the northern lights with magnetic disturbances, and to investigate the connection between temperature and rainfall, in which he identifed the concept now termed the dew point. Dalton finally moved from his village to Manchester in 1793, where he found that most (though not all) of his early ideas had already been anticipated by researchers with access to books, instruments, and each other. But his contemporaries recognized his talent, and within the year, Dalton won election to Manchester's Literary and Philosophical Society.

Joining the society launched Dalton into the critical phase of his work. His interest in how the atmosphere worked led him to ask how the different gases in ordinary air mix with each other and with other compounds. Lavoisier, among others, had proposed that that the oxygen and nitrogen that make up the bulk of the atmosphere bind together in some loosely connected form of a chemical compound. Dalton disagreed, arguing that the atmosphere is made up of a mixture of the two gases bouncing around in close physical proximity but remaining distinct chemically. He observed that different amounts of nitrogen and oxygen would dissolve in the same quantity of water (regardless of the presence of the other gas) and concluded that each retained its individuality in the atmosphere.

That observation propelled Dalton into his critical research. If different gases display different properties, he asked, what do those properties suggest about the fundamental makeup of those gases? Dalton responded that if every gas were composed of atoms each fundamentally different from each other, the distinctions in their size and weight would account for variations in the behavior of elements of matter at scales accessible to human senses. Assuming the atom, then, Dalton proposed a theory that described how chemical compounds form—and provided a line of inquiry that could confirm or deny the existence of atoms as the ultimate constituents of matter. He argued that all chemical compounds were composed of individual atoms of the different elements within the compound, forming a fixed proportion of one to the other, a ratio that remains constant for every molecule of the substance. Just as every atom of oxygen must be (according to Dalton's atomism) the same as every other oxygen atom, and different from the atoms of all other elements, so each particle of a compound like water would be identical to every other particle of water, and unlike any other compound.

He further suggested that elements formed compounds in the simplest possible way. If two elements combine, the most likely arrangement of the resulting compound would be, Dalton argued, one atom of each: A + B, in the language of Lavoisier's equations. The next most likely would be 2A + B, or A + 2B—and so on. Whatever the precise rules that determine the amounts of particular combinations, the underlying point was this: the ratio of one element to another in every compound will be an unchanging whole number. This idea, though it seems obvious now, was revolutionary at the time. By asserting that atoms form the discrete unit of chemical combination, Dalton created a simple, quantitative description of the composition of complex substances. Chemical analysis finally had a clearly defined goal: to count the number of atoms present in each compound. Moreover, Dalton's atomism enabled chemists to begin the systematic study of how differences in chemical composition could produce the different properties of even closely related compounds—to consider how carbon monoxide, for example, an atom of carbon bound to one of oxygen, behaves, for example, compared to its near twin, carbon dioxide, one carbon atom linked to two oxygen atoms.

Dalton tested this law of multiple proportions, as it was called, through such comparisons. Independent confirmation solid enough to be accepted broadly came in 1808 with the demonstration that one of two complicated potassium compounds contained precisely double the potassium of the other. These findings served as virtual proof of the existence of atoms: the fact that compounds consistently formed in whole-number ratios suggested as strongly as inference permitted that, at bottom, those whole numbers counted individual, indivisible units, bound together but not blended, not merged into some distinct, third substance. As a final proof of the existence of atoms as the fundamental form of all matter Dalton proposed one more test. He held that the atoms of each element differed from each other by weight. If atoms combine in whole-number ratios, then, he reasoned, it should be possible to construct a table of the relative weights of individual atoms based on the comparisons of the weights of the elements that make up each known compound. If those weights could be found, and shown to hold up across a wide range of compounds in which any given element was found, then there would exist a single number that could label each atom of each element. Describing the unseen (even by just the one quality that he could measure) constitutes discovery; if Dalton could weigh the atom, he could claim it.

And he could, sort of. Dalton began by attempting to relate the weights of hydrogen and oxygen, the constituents in water. Following his rule of thumb, that the most common form of chemical combination is the simplest, one atom + one atom, he assumed that a molecule of water consisted of a single hydrogen atom and a single oxygen atom. From there, Dalton had only to follow Lavoisier's procedure: weigh accurately a quantity of water, and then weigh the hydrogen and the oxygen that could be derived from it. He found that water yields eight ounces (say) of oxygen for every one of hydrogen. Dalton set the atomic weight of hydrogen at one unit, and thus assigned oxygen an atomic weight of eight.

That number is wrong, of course, and some of Dalton's contemporaries suspected the error. Looking at volumes rather than weights of gas, Sir Humphrey Davy, among others, noted that two cubic feet of hydrogen would combine with one of oxygen to form water. Making the assumption that there are the same number of particles of each gas in the same amount of space (which Dalton rejected), Davy suggested that molecules of water had to consist of two hydrogen atoms for each oxygen—as they do. Dalton's number was thus off by half: an oxygen atom actually weighs (approximately) sixteen times as much as a hydrogen atom. The error illustrates the single greatest hole in Dalton's theory: though it led to a method of determining atomic weights—loading up a scale with each of the components that can be assembled into or disassembled out of some chemical compound—the atomic hypothesis suggested no direct analytical procedure that could reveal the number of atoms of each element that went into any molecule under study.

Dalton recognized the problem, and he knew how to solve it.

177

He wrote that when trying to choose between different measurements of atomic weight derived from the study of different compounds, both containing nitrogen and oxygen, as one example, "we have to examine not only the compositions and decompositions of these two elements, but also compounds which each of them forms with other bodies." He had no illusions about his success: "I am not satisfied on this head," Dalton wrote in 1827, two decades after he first proposed his atomic theory, "either by my own labor or that of others, chiefly through want of an accurate knowledge of combining proportions."

Dalton's inability to come up with definitive figures for the weight of his proposed atoms left skeptics a loophole; some chemists continued to object to the whole idea of atoms as late as 1860—and some physicists, most notably Ernst Mach, denied the messianic vision of fundamental particles even later. But even without clear measures of atomic weights, most of Dalton's contemporaries, persuaded by the repeated confirmation of the law of multiple proportions, recognized Dalton's atomism as indispensable. Davy, once an antiatomist, wrote: "Let the merit of discovery be bestowed wherever it is due ... Mr Dalton's permanent reputation will rest upon his having discovered a single principle, universally applicable to the facts of chemistry. . . . His merits in this respect resemble those of Kepler in astronomy." Others rated Dalton even more highly. J. Liebig wrote after Dalton's death: "We, who stand in the presence of the science as now constituted, can scarcely conceive how it would have developed without this hypothesis. All our ideas are so interwoven with the Daltonian theory that we cannot transpose ourselves into the time when it did not exist."

The discovery to which Dalton can lay claim, that is, was not that of the atom itself—it lay beyond the reach of his skill as an experimentalist. Instead, as his fellow scientists recognized, he holds title to the invention of a principle, an idea. But Davy and Liebig, immersed in a world of science defined by Daltonian atomism, missed his greatest advance. Lucretius had suggested that the realm of atoms could be reached through reason, until their behavior and their nature could be identified, though the particles themselves remained ever remote from the direct measure of human senses. Dalton succeeded in proving him right. His theory *required* the existence of atoms. With sufficient precision over enough measurements, the atoms Dalton's ideas demanded could be—though he was not quite the man to do

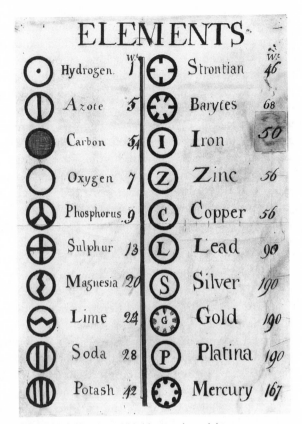

Dalton's table of the elements, with his atomic weights.

it—tracked down to their specific, individual characteristics, weight, and perhaps (Dalton speculated) eventually size and shape. At Dalton's urging, the invisible became (in theory) accountable.

All that remained was to turn theory into practice: to develop a comprehensive program of experiments that could establish reliable atomic weights—and hence, finally, confirm the atomic structure of matter. Doing so required above all a fascination with precision. The essence of any experiment is the measurement itself: counting whatever unknown quantity the experiment seeks to isolate. Fixing the weight of an atom involved, as Dalton noted, comparing the values derived from compound after compound. Any error in separating or combining elements would ripple through the entire series of experiments, destroying them. To catch a glimpse of any single atom—to account its weight properly—required, therefore, a sustained perfectionism, an obsession with accuracy, which Dalton himself could not muster.

The man who could was a virtuoso, a master of experimental technique. The procedures Sweden's Jöns Jakob Berzelius invented to track down the atom underlie all subsequent studies of atomic weight. His own application of them resolved the underlying issue unequivocally: as far as he and his contemporaries could see, the world, all of matter, was composed of many different varieties of discrete chunks of stuff. The world Berzelius confronted was complicated but intelligible.

Berzelius, born in 1779, seems to have determined on a scientific career early. Like Dalton, he was a poor child, and he began teaching at age twelve, taking on private tutoring clients on being sent to boarding school. Poverty later interrupted his medical studies, leading to apprenticeships with a pharmacist, and then a health-spa doctor. In fact, Berzelius's career as a professional chemist began when he identified the minerals in water at the Medevi spa in the summer of 1800. Strikingly, his analysis included a measure of the amounts of different compounds present rather than a simple list of contents. A generation younger than both Dalton and Lavoisier, Berzelius seems to have brought to his science from the outset a sense of the importance of specifically quantitative experiments. The older scientists broke trail for him—Berzelius simply accepted what they had had to prove, that "the use of the balance and . . . the balance sheet," as the historian Harold Hartley put it, formed a royal road to truth in studies of the material universe.

Berzelius weathered his early poverty sufficiently to complete the requirements for his medical degree in 1800. For the next several years, he took on medical posts, performing chemical research on the side. His early work reflected his medical background, focusing more on problems that he called "animal chemistry"—biochemistry, in contemporary jargon—than on the analysis of fundamental particles of matter. But on receiving a teaching job exclusively devoted to chemistry in 1807, Berzelius confronted the problem of teaching students the basics of the science as a whole. There was, however, no Swedish language textbook that covered the profound changes in chemistry that had taken place over the last several decades. To fill the gap, he decided to write one of his own, a chore that led him into direct confrontation with the atom.

Berzelius tripped over the hint that led him to the atom almost as soon as he began reviewing the chemical literature he had to summarize for his proposed text. A German chemist, J. B. Richter, studying the reactions involving acids and certain oxides of metal, noted that different amounts of different metallic oxides could neutralize the same amount of acid. Richter assumed that the amount of oxygen reacting with the acid would be the same. If that were true, then the amount of the base metal in each compound—monoxide, dioxide, trioxide—would have to vary in precise, whole-number proportions: 2:1, 3:1, 3:2. Richter apparently saw his results as simply successful experimental measurements—new knowledge, more precise information. To Berzelius, already convinced of the value of precise, quantifiable experiments, Richter's findings suggested a whole new field of scientific inquiry: with just a few measurements on the different oxides of a given element, Berzelius realized, he could work out the ratio of oxygen to element in all of the oxygen compounds of a given metal.

As Berzelius began to redo Richter's experiments, a recent number of an *Nicholson's Journal* reached him. *Nicholson's Journal* was an English periodical devoted to intellectual reporting, and the November 1808 issue contained a description of Dalton's atomic theory and the law of multiple proportions. He did not receive the full explication of those ideas until Dalton himself sent Berzelius a copy of his book in 1812, but that first account served as the starting gun. Berzelius recognized that, even in its compressed form, Dalton's theory combined with Richter's findings, meant that Richter's perfect ratios referred not just to the bulk weight of oxygen and metal in a compound but to the number of atoms that composed the molecules of each different oxidized form of an element. With that, the precise weights of the atoms within those compounds would reveal themselves in experiments exactly like those Dalton had already attempted.

Berzelius, though acknowledging the elegance of Dalton's theory, had little patience for the Englishman's attempts at experimental science. He wrote: "His discrepencies from the truth rightly surprise me; one sees how he seeks to model nature to fit his hypothesis. . . . As for the experiments which are his own many seem to have been well done, but one cannot trust them, as one sees always that the author has a preconceived idea, and I know only too well from my own experience how often one is moved to determine in advance the result of an experiment by which one hopes to prove some theory!" Berzelius scorned Dalton's habit of seeking the "right" answer in favor of a more subtle goal: designing a valid experiment. His acceptance of the atomic theory meant his measurements had to reveal atoms that combined in whole-number proportions. Given that as the test, Berzelius could judge the accuracy of his experiments by seeing how closely his figures for the weight of an atom of the same element found in different compounds matched each other. If the weights varied wildly, that result pointed to only one of two conclusions: either a particular experiment had been performed badly, and its answer was simply wrong; or the count of the number of atoms in one or another compound tested had to be in error, and needed to be recalculated. As Berzelius used it, the atomic theory served as the corrective and the judge of the quality of any single measurement he performed. Berzelius ran each experiment with no particular results in mind (Dalton's mistake, according to Berzelius), looking instead for the type of answer that had to exist, if the theory were true.

Berzelius justified his faith by measuring atomic weights more accurately than anyone before him—and demonstrating that the weights he found met the test of forming perfect proportions in the compounds he identified. The basic design of his experiments was simple, essentially the same as Lavoisier's on the role of oxygen and Dalton's first attempts to gauge atomic weights. Most commonly, he would take the oxide of an element and reduce it, removing the oxygen, while checking the weight of the compound before and after. The trick, of course, was to identify how many atoms of element combined with how many atoms of oxygen—which forced Berzelius to test several different ratios to track each atom to its place in the compound. Simple enough in conception, the actual analyses turned out to be extraordinarily complex.

For example, to determine the weight of the chlorine atom relative to oxygen, Berzelius studied two types of compound at once, combinations of chlorine with potassium, and with silver. To extract a value for the weight of chlorine from molecules which contained both potassium and oxygen, Berzelius began by boiling potassium chlorate—only to find that when the oxygen boiled off, it carried some of the original compound with it. Such inaccuracy was unacceptable, so Berzelius developed a method of heating the compound without boiling it. When the compound was completely dry, Berzelius would weigh it, and then attach a curved tube to the retort containing the chemicals being analyzed. The retort would be heated in a sand bath, providing dry

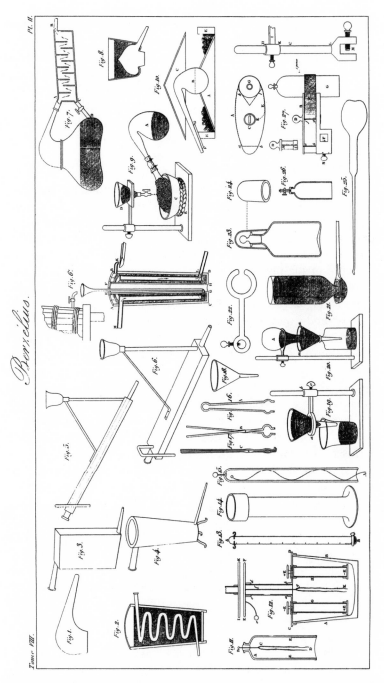

Part of Berzelius's apparatus for purifying elements and measuring atomic weights.

heat that would turn the compound red and soften it. Filter paper at the opening of the tube would catch particles of the compound that wafted upward by any smoke, while letting oxygen pass through into its closed end. When the sample in the retort ceased to give off any more gas, Berzelius would cap the tube, and then weigh it to measure the amount of oxygen present in the original compound. He would then let atmospheric air refill the retort, weigh that, and make sure that all the original matter in the experiment was accounted for.

In four measurements, Berzelius found that the weight of the tube increased between .02 and .023 gram; subtracting that from the weight of the original compound, Berzelius fixed the proportion of oxygen within his samples at between 39.146 and 39.150 percent—an agreement across four experiments of one part in ten thousand. This was, however, only the first step—and as a measure of the obsession required to do this kind of chemistry, recall that just to get this far Berzelius had to identify a method of removing oxygen and only oxygen from his test sample, which took several experiments; he then had to go through several steps to prepare the sample for its first weighing; he then had to weigh it; he then had to heat it, within a carefully constructed, closed system of retorts and tubes, all the while ensuring each piece of apparatus was sufficiently clean to allow him to collect and weigh quantities of a thousandth of a gram and smaller. All that done, the next task was to separate the potassium from the chlorine, in preparation for the final measurements that would allow Berzelius to determine the number of chlorine and the number of potassium atoms in the original compound—and thus finally determine chlorine's atomic weight. To extract the potassium from the chlorine, he dissolved potassium chloride with silver nitrate, and found that 100 parts of the potassium compound released enough chlorine to make 192.4 parts of silver chloride. That result, though, simply pointed to another measurement: now Berzelius had to identify the ratio of silver to chlorine in the silver chloride to help him fix the likely ratio of potassium to chlorine in the compound he had just disassembled. To do that, Berzelius needed to find out how much silver chloride can be produced using a given amount of pure silver.

That measurement turned out to be as complex as the original one to determine the proportion of oxygen in the first potassium chlorate compound. Berzelius began by dissolving silver in nitric acid. Then he evaporated the liquid to get rid of the extra acid;

then he dissolved the remaining silver nitrate again, this time in distilled water; then he dissolved the whole sample again in sal ammoniac, which contains chlorine. Then he passed the last solution through a filter, collecting the stuff left behind. Then, he weighed that dry precipitate to find, finally, that 100 parts of silver would produce 132.7 parts of silver choride by weight. This procedure can only err on the low side, since the filter might miss some material but could not gather what was not there. Just to be sure, though, Berzelius repeated the entire measurement using a different technique that could only miss on the high side. Those experiments yielded results of 132.78 and 132.79 parts by weight of silver chloride for every 100 parts of pure silver metal—close enough to the first number to give Berzelius confidence in the outcome.

But he was not finished yet. To determine the weight of the chlorine atom, Berzelius had to go through the whole series of exhaustive procedures to find the ratio of chlorine to oxygen in chloric acid; the ratio of potassium to oxygen within potash; the ratio of chloric acid to potash in potassium chlorate; the ratio of potassium to chlorine in potassium chloride; the ratio of silver to oxygen in silver oxide; and the ratio of chloric acid to silver in silver chloride (twice)—at least seven different, complicated sequences of experimental manipulation to produce quantities of chemicals that could, in the end, be weighed on a precision balance and related to each other with an accuracy of one part in 10,000. Only after all that, did Berzelius find his answer: an atom of chlorine weighed 2.2136 times as much as an atom of oxygen—35.46 to oxygen's 16, one one-hundredth of a unit above the modern value.

With that, and the similar measurements he made to fix the atomic weights of over 90 percent of the elements known at that time, Berzelius may fairly be said to be the man who discovered the atom. Many others—Dalton first among them—"knew" that atoms existed; only Berzelius developed the analytical skill and the commitment to precision in the laboratory to capture the atom in the pans of his balance. With his identification of specific, distinct numbers for the weights of each of the atoms for the forty-five elements he mastered, Berzelius established beyond the chemists' doubt that matter in its simplest form, at the level at which it could no longer be subdivided, was composed of a number of irreducibly different kinds of material particles. The differences between atoms, Berzelius and others suggested, would explain how different elements interact—how matter takes its various shapes, building the macroscopic world of human experience. Berzelius himself pursued that end by continuing his development of ever more sophisticated techniques of chemical measurement. Introducing his own results in 1818, he wrote that his goal was "not to obtain results which are absolutely exact, which I consider as only to be obtained by accident, but to approach as near accuracy as chemical analysis can go."

Thus speaks a genuine heir to Newton. Exactitude comes by accident, but by design the chemist can approach perfection. No measurement will be absolutely correct, but with labor and skill—Berzelius's hallmarks—one could surround the right answer with an ever tightening ring of increasingly certain knowledge. And yet, with the advantage of hindsight, Berzelius's statement reveals the minutest of fractures in the Newtonian worldview, the one that would deepen until the entire image was transformed. The fissure forms around Berzelius's acknowledgment of the inevitablity of error—though Berzelius himself did not sense its ultimate implications. For the most part, in fact, he simply expressed his confidence that the progress of science would lead to an ever more correct description of nature. And for a time, that confidence seemed justified.

With the atom in hand, Newton's hope of learning the "whole nature of bodies" seemed accessible to the further improvement of nineteenth-century science. The motions and forces that had been too small to grasp a century before were now being measured—leading to the belief that it would be possible to construct what Newton had proposed: a complete description of matter in the universe, accurate to an ever increasing level of precision, "as near accuracy" as one could go. The French astronomer Pierre Simon de Laplace was more overtly confident than most, but he caught the prevailing belief of his day when he wrote in the early nineteenth century that the revolution in science would eventually "embrace in the same formula the movements of the greatest bodies of the universe and those of the lightest atom." Given a calculating engine good enough to handle the numbers, "nothing would be uncertain, and the future as the past, would be present to its eyes."

Laplace is also the man who said, in response to Napoleon's question as to why he failed to mention God in his study of astronomical motion: "I have no need of that hypothesis." And he did not, although his friend and colleague Lagrange responded, wistfully: "Ah, but that is a fine hypothesis. It explains so many things." Ruthlessly, Laplace's science, with its absolute certainty and its pursuit of omniscience, took on the attributes of divinity, shouldering aside the God to whom all things are known, all hidden things revealed. The sense of expanding human understanding and control over the material universe was contagious. It is impossible to characterize an attitude of an era, as if

The Archangel Michael weighing souls in a detail from Hans Memling's *The Last Judgment*—an instance of a divine measurement Laplace did not attempt to duplicate.

any moment in history manifested a singular worldview. But certainly the dominant and dominating culture of the industrialized West in the nineteenth century believed that with the aid of a science that was clearly able to produce sure, unequivocal, correct answers, the world could be molded exactly to specifications.

Charles Dickens dissented, along with many others, loathing the complacency and the soullessness of the idea that one has only to measure in order to know. But Dickens's derisive characterizations express the prevailing view as powerfully as any of its supporters. Thomas Gradgrind, the grim voice of reality in *Hard Times*, arrives "with a rule and a pair of scales, and the multiplication table always in his pocket, sir, ready to weigh and measure every parcel of human nature, and tell you exactly what it comes to." Gradgrind gets his comeuppance quickly enough— all of his calculations fail to predict, for example, the course of his daughter's miserable marriage. But the belief in ever increasing power that Joseph Priestley had preached, years before Gradgrind made his fictional appearance, remained a touchstone. Sir Arthur Conan Doyle, with Dickens among the most popular writers of the Victorian age, was at greater ease than Dickens with the prospect of scientific perfectability. In *A Study in Scarlet*, his Sherlock Holmes rigorously upheld the potential omniscience of human reason: "From a drop of water," Holmes wrote in a magazine Dr. Watson happened to read at the very beginning of the famous partnership, "a logician can infer the possibility of an Atlantic or a Niagara without having seen or heard of one or the other. So all life is a great chain, the nature of which is known whenever we are shown a single link of it."

A Study in Scarlet, with its tale of a complex and obscure crime reduced to "a chain of logical sequences without break or flaw," was published in 1887. At the same time, as the historian J. D. Bernal put it: "The increasingly coherent and unitary picture of the sciences that ... progress in the nineteenth century had revealed seemed to the scientists a sign that science was nearing its end." That end was the one Laplace had glimpsed: Berzelius's modest-seeming claim that his analysis simply provided the best available account of the atom generated the grand vision of a universe entirely composed of atoms, accessible to investigation that Laplace had glimpsed. James Clerk Maxwell, the greatest physicist of the nineteenth century, wrote of atoms "unbroken and unworn" that "continue this day as they were created—perfect in number and measure and weight." All that was left to do was to count them up, put them on the scale, and click a stopwatch to gauge how fast they travel. With that, the answer would pop out: this is how the universe behaves; this is what it is and will be. At the end of the century, it appeared that no great outstanding secrets remained, justifying what Bernal described as "a picture of fate more all-inclusive than any the Greeks had had"; only the details were left.

And yet, by the last years of the century, that inclusive vision was on the verge of disintegration. The atom itself, revealed by Berzelius's analysis as the immutable, eternal, and fundamental particle of matter, was about to be transformed, opening anew the old question of what the world is made of. In 1897, the first subatomic particle, the electron, was discovered, revealing structure behind the smooth facade of the chemical atom. The discovery of the proton followed, and then the phenomenon of radioactivity shattered one of the essential principles of the old atomism, the permanence of atoms. Radioactivity succeeded where the alchemists had failed, genuinely transmuting elements, converting atoms of uranium, for example, into lead (but sadly for the millennia of alchemical hopes, not into gold).

The advances in atomic physics did improve the precision of measurement that was Berzelius's goal. The mass spectrometer, invented in 1919 by F. W. Aston, served as a kind of electromagnetic scale, replacing the two-pan balance after three thousand years of service. It uses a magnetic field to deflect particles—atoms or molecules—of the same charge but different weights. Differences in the amount of deflection between two such compounds give a very precise value of the relative weight of each particle. While Berzelius, along with all of his contemporaries, assumed that there was only a single type of atom for each element, the mass spectrometer revealed the existence of atomic isotopes—different forms of chemically identical atoms. Oxygen, for example, comes in two main forms: ^{16}O and ^{18}O—the latter containing two extra neutrons (the third subatomic particle to be discovered). ^{18}O thus weighs 11 percent more than the more common ^{16}O form. The oxygen that Berzelius weighed and used as his reference standard was actually a mixture of the two isotopes—but Aston's mass spectrometer separated the lighter atoms from the heavier. With that he was able to adjust the weight of the ^{16}O atom down just a little—accounting, among other things, for the revision of Berzelius's determination of the weight of the chlorine atom, expressed in terms of oxygen.

This was exactly the kind of triumph of measurement Berzelius had in mind, as he aimed not for the exact answer but for the best possible answer. But it came too late—or rather, it came amid the growing realization that however well one weighed each particle of the universe, any ultimate description of nature was unattainable. The atom the chemists invented was hard, mechanical, indivisible—a kind of miniature marble. The atom the physicists depicted in the strange new mathematical language of quantum mechanics was a convention, almost a fiction, composed at once of both particles and waves, visible in glimpses but never to be seen whole, front and back. Heisenberg's uncertainty principle, uncovered in 1927, emerged directly out of the equations of quantum theory, and it drove the final nail into the coffin of Newton's hopes.

Heisenberg realized that any measurement affects the quantity being measured. That does not matter much if the sample is large enough, but it means that no atom ever gives up all its secrets. Specifically, Heisenberg noted that it was impossible to measure absolutely accurately both the position and the velocity of a sufficiently small moving particle in any single experiment. Other pairs of qualities exist as well, where the measurement of one property limits the accuracy with which the other property can be known. The bottom line is that each measurement must exclude some information to increase the precision with which other data can be gathered—the more one knows about where a particle is at any moment in time, the less can be said about how fast it was moving to get there, or how fast it will take off when one's back is turned. With that loss of certainty, Newton's vision (as dreamed by Laplace) of perfect knowledge of the material universe, of both past and future laid bare, could not be attained by any human art.

But long before the quantum theory finished off the earlier revolution that had been the product of Newton's waking thoughts, the search for the atom had already defined limits to the power of the science that had emerged between the seventeenth and nineteenth centuries. Berzelius's grasp of the idea of the best possible measurement compelled him to accept Newton's belief that one could describe the universe with a set of mathematical laws. To recognize the "best" outcome, that is, Berzelius needed to know that there was a "right" answer, one both predicted by law and generated in nature. Berzelius clearly acknowledged that somewhere in the gap between best and right lay a limit to his ability to gather every fact about nature. Techniques of measurement could be improved to the point of perfection—but only to that point and no further.

With that, the Newtonian goal to know everything evolved into a narrower ambition: to know everything that can be known. To Berzelius and to contemporaries like Laplace, the distinction did not seem to matter much. Every measurement still added to the picture of nature, one that remained slightly out of focus, perhaps, but essentially correct. Improved technique would sharpen the focus but leave the image itself unchanged. The errors that would remain seemed sure to be trivial. And yet, the last great discovery of the scientific revolution (well before quantum theory emerged) was the realization that those errors count.

Some—not scientists—recognized that what could be measured did not necessarily tell the whole story. There is a story from the history of music that illuminates the course that the history of science took, one in which the making of a musical instrument reveals the limits to what can be learned by the use of scientific instruments. The emergence of the violoncello in its modern form took place in 1708, in the workshop of Antonio Stradivari in Cremona Italy. Yet though we know everything that can be counted about Stradivarius's instruments, we cannot build their like, even now. The central tenet of scientific faith of the nineteenth century was that the methods scientists used (theories made up of mathematical laws; experiments based on precision measurement) actually described nature as it is. That belief retains force still, but it totters, undermined by the realization that science describes not nature but what can be measured within nature—and that measurements are not enough: Stradivarius has no equal.

Thus, having failed to learn "the whole nature of bodies," the final task Newton's inheritors faced was to define exactly what their science could do, what any scientific experiment actually reveals about the universe at large—and what remains hidden. After two hundred years of routine scientific triumph, such questions struck a dissonant chord—but actually, the mystery they confront is as old as the attempt to make sense of our surroundings. "We must not ask nature to accommodate herself to what might seem to us the best disposition and order," warned Galileo Galilei, long before this story played out, "but we must adapt our intellect to what she has made, certain that such is the best, and not something else."

The discovery that not all that we might wish to know is knowable seemed tragic, unthinkable, to many who witnessed it being made—those who could not adapt their intellects to this unwonted face of nature. The victors in the scientific revolution had sought perfect understanding, and with it the dominion of humankind over the natural world. What they gained instead was a critical leaven of self-knowledge: there is a finite bound to the power of scientific inquiry, to the reach of that human aspiration that would capture the universe whole.

Chapter 6

THE
EXACTITUDE
OF THEIR
PROPORTIONS

It is only a coincidence—or rather, mostly so—that Isaac Newton, Johann Sebastian Bach, and Antonio Stradivari reached the pinnacles of their professions at almost exactly the same moment in history. Their lives and work did not touch, though at times they almost intersected. Newton's fascination with the tempered scale overlapped with Bach's. Bach's omniverous musical interests extended to the labor of instrument design and construction that was Stradivari's lifework. And Stradivari (also known by the Latin form of his name, Stradivarius), in the course of his seventy years as a master luthier and builder of violins, violas, and cellos, applied a craftsman's version of Newton's scientific method, deriving by experiment the skills needed to create the finest stringed instruments ever built.

But the three men never met, and if their lives show some common heritage, it is because the ideas that moved them moved their times, forming a common currency of inquiry and change across Europe. The real collision among the three of them came much later. Taken as a drama, it is Stradivari who plays the lead. His is a story of mystery: the core principle of the new experimental science was that what one person can discover any other can find again—yet for three centuries none have been able to do what Stradivari did, to build an instrument that reproduces the range, power, and expressiveness of the Stradivari sound. Here Newton enters, or Newtonians, or, better, the claim of Newtonian science that the universe is knowable, totally. As the attempt to isolate the secrets of Stradivari's accomplishment proceeded over the years, the unmistakable individual character of his instruments became the embodiment, the physical evidence of a heresy: there are, there could be, strictly defined limits to the amount of precision with which Newton's scientific apparatus can account for the material world. As far as cello and violin making goes, Stradivari did not merely use Newton's methods, he exhausted them.

Which is to say that Stradivari's story and the tale of how he developed what became the modern form of the cello provides a case study of an application of the scientific method. But the history of the cello, taken more or less in isolation, leaves out the larger implications of a scientific failure to comprehend how one form of musical sound can be made out of the simple raw materials of wood, gut, glue, oil, gum, and air. Here enters the third man, Bach, whose work offers a commentary, an epitaph and an insight into the process of trying to make sense of any phenomenon. Though the central focus of what follows is the physical problem of how one constructs an instrument to perform a task, there is a point of entry into that story through what was perhaps Bach's most abstract, most purely reasoned work, his late, great composition, *The Art of Fugue*.

In Bach's day, and in that last effort most of all, music retained its kinship with science, with the conscious exercise of rational thought, that it had possessed since Pythagoras. Music, like science, music as part of the same endeavor as science, was a tool that could both describe and discover patterns, order in nature, in human experience. The mathematical precision of rhythm and harmony suggested metaphors and the real thing, all at once—hence Kepler's polyphonic solar system and Newton's rainbow, mapped onto a scale of colors. But in Bach's lifetime, and in part at his hands, music's relationship to science, music's status as *a* science, had already been undermined. Bach composed *The Art of Fugue* between 1745 and 1750. It contains twenty-one canons and fugues; the last, a fugue with one subject built on the notes that spell Bach's name, is unfinished. (The theme goes B-flat, A, C, B-natural—in German notation B means B-flat and H means B-natural.)

In its place in the history of music, *The Art of Fugue* forms a monument to the art of counterpoint, to the concept of polyphony—the multivoiced music invented in the Middle Ages. Bach inverts themes, constructs mirror images of themes, intertwines multiple voices to produce increasingly complex harmonies—and yet preserves each voice, each melody, each theme to reach its own conclusion at the end of every piece. Technically it is a masterful composition. Bach's skill enabled him to manipulate musical ideas so that they both blended harmoniously and retained their clear, independent coherence—and it was the virtuosity with which he did so in *The Art of Fugue* that capped the golden period in the history of counterpoint. Music after Bach shifted its focus from the interplay of distinct voices, counterpoint, to one that depends more purely on a single melody, resting on a bed of harmony. Harmony in this sense refers to the construction of a musical sound out of the mixing of notes together in a chord (replacing the older notion implied by the phrase "the harmony of the spheres," which referred to qualities like balance and order). The conversation of counterpoint, that is, can be compared to the chorus of harmoniously blended music. That harmony moves from one chord to the next, through consonance and dissonance, to create its musical effects. The shift from counterpoint to this style of melody-plus-harmony was a move that musicians refer to as a change from horizontal music—voices that alter and intertwine over time—to vertical music: harmonies built by placing note upon note to create a stack of sound, all heard at one moment in time. Compare Beethoven to Bach to hear the difference. Beethoven (like Bach) had a gift for orchestration, of course, an ability to construct astounding chords built out of the different sounds his orchestras offered him. But even more, his stock in trade was his revolutionary ability to stretch the limits of harmony—to create combinations that sound ever more precarious, ever more on the verge of pure noise—and then to resolve them, to bring them to a consonant, resounding conclusion.

The change from one musical idea to the next was not absolute, of course. Just as the scientific revolution retained and transformed older modes of thought and beliefs, composers before Bach had begun to explore harmonic possibilities, and many after him did write counterpoint—in fact, there was a resurgence of interest in the style in Beethoven's own time. There are fugues from the nineteenth century that are more formally complex than any Bach ever wrote—but they do not sound particularly pleasing, and they are rarely if ever performed. Beethoven, as it happened, like Brahms who followed him, sometimes found counterpoint difficult, though many of both men's works display great contrapuntal virtuosity. Bach himself lived long enough to see his music go out of fashion. He died in 1750, blind and out of step with the music that surrounded him, a style often called *galant*. His son, Carl Phillip Emmanuel Bach, published *The Art of Fugue* in 1752, commissioning a preface by Bach's colleague Friedrich Wilhelm Marpurg. Marpurg condemned the age that would ignore the master: "A melody which agrees only with canons of taste obtaining at a particular time and place," he wrote, "has value only so long as that taste prevails." (Marpurg also left no doubt what he thought of contemporary preferences, complaining of the "hoppity melodification" that surrounded him.)

Despite Marpurg's best efforts, Bach's last work was at first largely ignored. C. P. E. Bach noted in 1756 that just thirty copies of *The Art of Fugue* had been bought. The original plates were broken up and sold for the value of the metal. The tale of Bach's eclipse is a famous one in music, and is often, like all good stories, overtold. The old master's work had in fact moved increasingly out of step with the desires of his audiences, especially the town fathers of Leipzig. Nonetheless, musicians and composers clearly understood the extent and power of Bach's achievement, and his reputation slowly gained ground as the decades passed. Beethoven himself made Bach one of his muses —as a twelve-year-old he studied *The Well-tempered Clavier*, the work Bach had originally written for the instruction of his own sons—and near the end of his life, he recorded in his journal that he had been pondering a portrait of the composer, who, with a handful of other heroes could "promote my capacity for endurance." But for all such admiration, and despite the Bach revivals that began in both Germany and Britain in the early nineteenth century, the shift that Marpurg deplored did actually take place around the time of Bach's death, and with it came an alteration in what seemed beautiful—and true—in music.

In a sense, Bach did himself in. Terms like *subjective* and *objective* have had most of their meaning beaten out of them, but they suggest something of what happened. *The Art of Fugue*'s intricate, formal, abstract structure shows Bach's use of his tools of reason, an effort directed toward the end of building patterns whose shape could be discerned by any listening ear. Any exercise in counterpoint, even those written today by apprentice composers, requires such reasoning, demands the production of such objectively identifiable patterns. But Bach reasoned better and more deeply, coming up with richer patterns than others had, and he housed within his formal structures "subjectively" beautiful musical figures. He did it better, and more elegantly, than anyone else. And because of that extraordinary ability, Bach's music retains in the rigor of its mathematical structures a sense of the absolute, of impartial, impersonal accuracy—of the truth that we discern when we can identify the patterns a phenomenon creates for us.

But with the shift from counterpoint to harmony came an equivalent switch from a music that could seem absolutely true (like the contrapuntal song of the planets) to one that seemed to state particular truths. Harmony is not absolute—a dissonant note in one chord can become the foundation of the consonant, resonant, resolving flourish of the next. Working with such ambiguity enabled Beethoven to create the striking, vertiginous journeys that could lead a listener to his conclusion of joy, exaltation, sorrow, or despair—the composer's truth made universal (or at least broadly intelligible) by the particular beauty, the singular power of a given composition. Crucially, that individual expressiveness emerged directly out of what had seemed to Bach's immediate heirs an almost pedantic love of the universal logic of counterpoint. Bach helped to prepare the ground for Beethoven and the others, in part by seeming to have done almost all that could be done with the older tradition (thus forcing them to look for something new). At the same time (as Beethoven, studying Bach's scores, well knew) he had pointed out the most promising avenue to explore. His well-tempered scale signaled of the end of the quest for ultimate perfection. Instead, it offered the giants of nineteenth-century music a musical scaffolding on which to assemble vertical stacks of notes that would have sounded impossibly harsh under older tuning systems. Bach's exercise of freedom provided a cue—one of many perhaps, but an extraordinarily powerful one—that those who came after him seized upon with a vengeance.

And here, the link between music and the science of the day was broken, for a time. Music as an expression of personal discovery had little to offer a discipline determined to identify the facts as they stood, all the facts. Musical experience became an object rather than an agent of scientific inquiry. The issues of how sounds are made, how they interact, what distinguishes what we hear as having a particular musical effect—a tone considered as harsh or sweet, for example—all these and more became simply entries in the long list of problems for physicists to solve. Solve them they did, in many instances—but not in all. This is where Stradivari's story begins, the tale of his (almost) miraculous cellos, providing one account of that lapse and a hint of why it occurred.

The story begins much more quietly, within a handful of workshops concerned simply with the difficulty of meeting a novel set of demands in the music business. The entire violin family—violin, viola, and cello—was a latecomer to the suite of "classical" orchestral instruments to emerge. The earliest violins had just three strings, and they show up in paintings from Milan in the first decade of the sixteenth century. The first true, four-string violins appear a few years later, with the rest of the family making their way on stage at about the same time—there is a fresco at Saronno Cathedral, dating from about 1535, that shows both a viola and a cello. Initially, there was just a whiff of the vulgar around such instruments. The violin itself evolved out of several fiddlelike Renaissance instruments, in particular the rebec and the lira da braccio (literally, the arm lyre), used most commonly for dancing or for accompanying singers. In its early years, the violin served much the same function: professional musicians employed them to provide entertainment-for-hire. Serious music for strings belonged to the members of the older, more distinguished viol family—instruments that use a bow on their strings, like the violins, but that have frets, like guitars. As the sixteenth-century writer (and snob) Jambe de Fer put it: "We call viols those with which gentlemen, merchants and other virtuous people pass their time. . . . The other type is called violin; it is commonly used for dancing."

Renaissance dance music did not put any extraordinary demands on the violin or the violinist, but as the sixteenth century wore on, the rich and the powerful began to cultivate their leisure—and the music with which they filled their easy hours. Beginning in about 1550, demand for high-quality musical instruments was great enough to support a growing number of master craftsmen: skilled workers whose product could command the highest prices. The first known great violin maker was Andrea Amati of Cremona. Along with his contemporary, Gaspar da Salò of Brescia, Amati began to create instruments with individual, distinct voices—far superior to the fiddles dancing teachers used to put their pupils through their paces. Such fine workmanship fed an international market. In 1588, da Salò complained in his tax documents that his export sales had slipped a bit, while Amati earlier had fulfilled the most impressive order

for musical instruments placed in the sixteenth century, building
a set of thirty-eight instruments (twenty-four violins, six violas,
and eight cellos) during the 1560s and 1570s for delivery to the
court of Charles IX, king of France.

With such patronage the prestige—and hence the popularity—of the violin grew to the point that by 1600, violin makers
had appeared in most of the major kingdoms of Europe. But the
craft was advancing fastest in a few towns in northern Italy,
especially Amati's Cremona, a small city southeast of Milan, near
the Po River. Amati was the patriarch of the Cremonese violin
trade, founding a family of violin makers that included his sons
Antonio and Girolamo. With his grandson Niccolò, the Amatis'
violin-making skills reached their peak. Niccolò was also the best
teacher of his day, training most of the Cremonese master builders of the seventeenth and eighteenth centuries.

By the time Niccolò reached the prime of his career, in the
middle 1600s, the Amati family, along with the other makers
clustered in Cremona, had managed to create instruments that
set the standard for musical performance at the time, violins that
possessed a recognized, highly valued combination of versatility
and beauty. Well-made Cremonese instruments produced (and
generate still) an excellent tone—bright and clear—and violins in
general were far more flexible, capable of producing different
types of sound for dramatic effect than the older, more cumbersome viols. Conservatives still complained: Thomas Mace wrote
in 1676 that violins were merely carping instruments, erupting in
a "High-Priz'd Noise fit to make a man's ear Glow, and fill his
brains full of frisks." ("Frisks" are freaks or whims.) But the
range of sound and feeling that the newer instruments could
express drove the spread of the violin family throughout musical
Europe. Violins and violas came to dominate the emerging field
of opera orchestras, and composers working in new forms like the
sonata and the concerto wrote an ever larger repertoire specifically for the upstart violin.

Up to this time, the latter half of the seventeenth century, the
production and refinement of the violin seemed to hold no fundamental mystery. Different makers experimented with instrument design, attempting to match their product to changing
patterns of use. In their hands the violin evolved, pushed forward
by makers who knew what they wanted to hear and how to get
it. Some early violins, for example, often possessed a sweet tone
but a small voice. The lack of power did not matter as long as

violins remained within the drawing room or a smallish dance hall. But with the growth in the popularity of the violin, the best builders began to explore the sonic potential of their instruments by experimenting with a variety of designs. Niccolò Amati, for example, altered the family pattern, building violins with larger bodies, capable of filling a large room from front to back, while retaining his family's prized excellence of tone and nuance—and other makers found other solutions, variations in the shapes of their instruments, to achieve similar results.

In their daily work, that is, the violin makers of the seventeenth century had as clear a sense of the connection between cause and effect as did any of their more formally learned contemporaries carrying on the scientific revolution. The habits of thought and experimental practice on either side were essentially the same: Niccolò clearly understood the link between the acoustics of resonating chambers (the violin body) and the amplitude—volume—of the sound one could produce with different resonators. Systematic trial produced an instrument that worked, using the technique suggested by Isaac Newton himself, according to his familiar inductive reasoning, the method of analysis that proceeded by experiment and observation. Presumably, knowledge and skill would continue to accumulate, generating a continued, steady, comprehensible improvement in violin design and construction leading ultimately to the Newtonian goal, the discovery of the "laws" of violin making.

Certainly, such a course of events seemed to hold for the violin's larger sibling, the cello. The cello, the bass member of the violin family, is pitched an octave and a fourth below the violin, with its lowest string usually tuned to the C two octaves below the piano's middle C. The early history of the cello was actually less promising than that of the violin. Like the smaller instrument, the cello suffered the social stigma of vulgar entertainment—it even endured the humiliation of having holes drilled in its back to accommodate the strap that allowed it to be used in marching bands and public processions. Solo cello music dates back to the 1620s, but the cello's primary job early on was to join with a keyboard instrument—harpsichord or organ—to provide the musical foundation for a piece while a singer's voice, a violin, or some other intsrument grabbed the soloist's glory. The combination of cello and keyboard was called the *continuo* in that it provided a continuing accompaniment for the solo line of a composition. The cello often performed this role in church, joining the organ to play the bass lines of a growing repertoire of religiously inspired chamber music. The need to produce a lot of sound to fill a church dictated the fundamental design choices for the early cellos; they had to be loud, hence large.

Thus, virtually all cellos built before 1680 were substantially larger than those in use today. The earliest surviving cellos date from the middle of the sixteenth century, all six of them made by Andrea Amati—including one, called "the King," that was probably part of Charles IX's set of instruments. Originally its body was very large by modern standards, more than thirty-one inches long, compared with the current usual length of about twenty-eight to thirty inches. ("The King" itself now measures $29\frac{11}{16}$ inches; it, like most of the old, oversize cellos, was cut down to the newer standard in the eighteenth and nineteenth centuries.) Such instruments were tuned a whole step lower than the smaller instruments coming into use; their lowest note was a B-flat, rather than a C. Pitch levels were lower too—the seventeenth century's B-flat would have been a lower frequency than the one we use today. While they lasted, cellos of this size produced a fine, powerful voice, strengthening especially the deeper notes of the instrument. But the price paid for volume was that such cellos were relatively difficult to play: the distance between the notes on the fingerboard meant that cellists could not perform with anything like the facility of violinists and violists.

The situation began to shift around the middle of the seventeenth century. Composers began to write more and more music—secular as well as religious—that specifically called for the church bass, the large cello, instead of the equivalent member of the viol family, the viol da gamba, until by the end of the century enough music existed to support a few full-time professional cellists. Those players in their turn sought opportunities to perform the same kind of solo parts that the violinists already enjoyed. Composers like Corelli were willing enough to oblige with music that emphasized strong cello lines, but the church version of the cello was not adequate to the task. The violin had evolved relatively slowly to meet the changing requirements of the music it was supposed to make. By the 1680s, such incremental change would not do for the cello; what was needed was a radical reconfiguration of the instrument.

Niccolò Amati died in 1684, just before these pressures forced the development of a new form for the instrument. But throughout his life Niccolò trained a series of apprentices, virtually all of whom, it seems, caught their master's willingness to experiment. In 1656, or thereabouts, Niccolò, then sixty years old, accepted one more boy into his workshop, a youth of twelve or a little older, whose work would culminate in what would become the definitive design for the modern cello. His name was Antonio Stradivari.

Very little is known about Stradivari's early years—there are no records of his birth, nor any mention in parish registers of baptism. The first hint of his existence comes from his work. A violin made in 1666 bears his label, and adds that the maker was an "alumnus Nicolai Amati"—one of Niccolò's students. (Stradivari actually made that customary acknowledgment of his teacher in only one surviving instrument—all his subsequent instruments, even those made while working for Amati, bear his name alone.) Lacking much in the way of conventional documentary evidence, Stradivari's biographers, Henry, Arthur, and Alfred Hill, reconstructed the first stages of his career based on the traditional sequence of an apprentice's training, combined with the initially scattered testimony of the instruments in which they could detect Stradivari's hand. A skilled beginner could produce complete instruments of his own three to five years after entering a master craftsman's shop, which means, according to the Hill brothers, that Stradivari probably made his first instruments around 1660. Certainly, by the middle of the decade, his work was prized enough to attract commissions of his own, independent of the orders coming in to the Amati concern. Nonetheless, Stradivari seems to have remained in Niccolò's workshop until 1680, when the older man probably retired at the age of eighty-four.

Up until that time, Stradivari built instruments that followed the basic Amati pattern, but once on his own, and then more rapidly after Niccolò's death, he accelerated his lifelong process of innovation. Much of the 1680s seems to have been spent refining craft skills: instruments from the end of the decade show the most precise carving, fitting, and workmanship Stradivari had achieved to that date; the Hill brothers point to the "Tuscan" violin, made in 1690, as an example of the finest craftsmanship Stradivari ever produced.

In the 1690s, though, Stradivari embarked on a rapid series of experiments. Seeking the power and depth Niccolò had aimed for with his large-model violins, Stadivari went one better, building six (known) instruments that are longer, wider, and occasionally deeper than anything else he ever built. The instruments did not work out as planned—they gained volume and strength in the lower notes at the cost of the bright and lively tone both the Amatis and Stradivari had sought. So Stradivari took the next step, producing a series of violins on what is now called the "long-strad" pattern: instruments that retained the length of his massive earlier version but were cut narrower and thinner, back to front, than previously. They played better than his previous attempt but still fell short of the sound he required. Finally, around the turn of the century, he began building shorter violins shaped to a flatter curve than that favored by the Amatis for the arching on the backs and bellies of the instruments. These became the classic "golden period" instruments—possessed of the combination of power and clarity of tone that created the Stradivari reputation as the finest violin maker in the history of the instrument. (His only rival in prestige and price, at least, was and remains Giuseppe Guarneri del Gesù, the grandson of another of Niccolò Amati's apprentices. Several great violinists, including Paganini and currently Isaac Stern, have preferred his instruments to Stradivari's.)

In contrast to his almost feverish pursuit of new ideas for his violins, however, Stradivari seems to have paid relatively little attention to the cello. He made several, beginning in 1680, but until 1701 he always followed the older style, producing large, loud instruments. Then, for six years, he ceased making any cellos at all. He left no explanation for the break, though the success of his violin experiments and the growing volume of his business may simply have distracted him. But in part, he seems to have spent the time surveying the landscape, considering the next direction to take. Beginning in the 1690s, a few Cremonese makers had begun to experiment with the cello form, coming up with different, smaller patterns in an effort to meet the demand of professional cellists seeking a more playable instrument than the older, so-called "basse di violon." Stradivari almost certainly came into contact with such prototype instruments, becoming familiar with the fundamental problem of preserving a big bass sound in the new form. In 1707, he produced his response, building the first of a series of instruments that set the standard form for all modern cellos. The cello of 1707 represented a breakthrough: less than thirty inches long, it was compact enough to permit a performance that was swift and precise without sacrificing its quality of sound. As the Hill brothers wrote: "The supreme merit of violoncellos of this type ... consists in the exactitude of [their] proportions. ... They stand alone in representing the exact dimensions necessary for the production of a standard of tone which combines the maximum of power with the utmost refinement of quality, leaving nothing to be desired: bright, full and crisp, yet free from any suspicion of either nasal or metallic tendency."

By the time Stradivari began producing his small-form cellos he was already recognized as one of the premier instrument builders in Europe, frequently receiving aristocratic, even royal commissions, back-ordered to the point that the wait for one of his instruments stretched over years (even for kings). Consequently, even while Stradivari was alive and working, his prices tended to be considerably more than a working musician with no private fortune could afford—and the gap separating the performer from the instrument has widened ever since. About fifty Stradivari cellos still exist, about half of which reached the perfection of form and sound the Hills praised. These instruments have become legendary—the stuff of romantic ardor and of the mythic aura that surrounds Stradivari to this day. During the nineteenth century in particular, Stadivari cellos seemed to mimic the (imagined) roles of the great divas of melodramatic fictions: they were desired but never truly possessed by rich men—who, for the sake of art, had to surrender them to their poor but honest lovers.

A cello made in 1714, for example, made its way to Madrid and from there to Paris in 1836. The French cellist Adrian François Servais, who played the last of Stradivari's large cellos, told his friend and fellow cellist Alexander Batta of its arrival at the instrument dealer's. At the time, Batta was perfectly happy with his recently purchased Amati cello, and in any event he could not afford the asking price on the new instrument. But at Servais's urging he stopped by the dealer's shop to compare his Amati with the Stradivari. "I saw it, played upon it and was completely captivated," Batta later recalled. "Never had I experienced such a passionate longing to possess an instrument." The price, 7,500 francs, was still beyond him, but the cello overcame Batta's pride. He turned to "a dear friend" who, moved by such exuberant desire, went round to the dealer's shop, bought the cello, and handed it over to Batta on the spot. Batta kept the instrument for the rest of his life, rejecting all offers, including that of a Russian noble who signed a blank check and told Batta to fill in the price.

Batta was prudent: at least as the Hill brothers saw it, his Stradivari was vital to him, "sharing in and largely contributing to his successful career." (Batta's cello now resides in the Library of Congress, and has been used by Anner Bylsma for both live performances and a number of recordings.)

Similar stories follow other Stradivari creations: the Piatti cello, made in 1720, gained its name from its most famous nineteenth-century owner, the great Italian cellist Alfredo Piatti. In 1867, Piatti paid one of his frequent visits to a friend in London, a Colonel Oliver, a collector of old cellos. At the time the colonel owned three—the Stradivari, an Amati, and an instrument made by the fine Venetian maker Domenico Montagnana. After Piatti played each, his host asked him which he preferred. "One cannot have a doubt," Piatti said. "The Stradivari." The musician left the collector and returned to his rooms—only to find that the prized Stradivari followed him home. Such gestures have continued to accompany Stradivari instruments, even in the late twentieth century. The Davidov cello of 1712 has three times been bought by wealthy amateurs and then given to brilliant professionals. First, Count Wielhorsky traded a Guaneri cello, 40,000 francs, and his best horse for the Stradivari, which he then offered to the leading Russian cellist of the day, Karl Davidov. In the 1960s, a wealthy benefactor bought the cello for Jacqueline Du Pré, the spectacular English cellist whose career was cut short by multiple sclerosis. Finally, when Yo Yo Ma, to whom Du Pré had loaned the cello after she ceased to perform, could not afford to purchase the instrument on his own, another rich music lover once again bought it (through a foundation) for Ma's use for the rest of his life.

The reason these cellos have had such colorful lives, of course, is because they produce a uniquely beautiful sound and range of expression. The best way to get some sense of what makes them so prized is to listen to them, at concerts if possible, otherwise on disk. For just two of many choices, Ma will be recording the Bach cello suites for the second time in his career in the mid 1990s. That will permit listeners to compare a Stadivari to another fine instrument, a cello built in 1733 by the Venetian maker Montagnana, that Ma used on his first recording of those works. Additionally, Mstislav Rostropovich's recordings of Benjamin Britten's solo cello works show off the extraordinary range of both the performer and his 1711 Stradivari.

But hearing what has made Stradivari instruments the objects of such desire merely begs the next question: what makes his cellos perform as they do, outplaying those of other makers? If we abandon the benefit of hindsight, Stradivari appears simply to have continued the steady progress of innovation and improvement begun 150 years earlier by Andrea Amati. Only in retrospect do his instruments appear to us to have reached a peak that no one before or since has managed to scale. Montagnana and Matteo Goffriller, among others, succeeded in building cellos that produce a beautiful sound. But in general, after Stradivari's death in 1737, the tradition of fine instrument building withered in Cremona, and the other centers of the craft in Italy never matched the extraordinary flowering that culminated in Stradivari's career. That is his true mystery: What did he do to climb that peak—and why have his successors been unable to match, much less surpass, that accomplishment?

The answer to the first part of that question—what Stradivari did as he made his instruments—is actually well known. The cello, after all, is simply a machine for making a particular set of sounds. As such, how any cello works, including Stradivari's, can be analyzed in detail. It, like all the members of the violin family, generates a note when a string is plucked or stroked by a bow, setting up a rapid vibration along the length of the string. Such vibration produces almost no sound in itself—one violin maker has compared it to the act of fanning the air with a toothpick. Instead, a carved arch of wood, called the bridge, that fits between the cello's top and its strings transmits that motion from the strings to the cello's top plate, known as its belly. The sides of the instrument—called ribs, as the anatomical metaphor persists—transfer the belly's oscillations down to the cello's back, as does the sound post, a piece of wood that is wedged (not glued or fixed) between the two plates. The bass bar, a long and thin strip of wood glued to the inside of the belly, alters the way the cello top responds to the pressure of the bridge, both strengthening the structure and shaping the tone of the instrument.

At that moment, with both the back and the belly in motion, the body of the cello acts as a resonating chamber, amplifying the original vibration—the note—produced by the stroke of a bow on the string. Critically, the bits of wood that create that resonant box do not all flex in a single, continuous motion. Doing so would produce a sound with most of the energy concentrated around a single wavelength, or pitch—what is called the fundamental. Flutes produce quite pure fundamental tones, as do Pythagorean monochords. But most instruments create a sound that includes both the fundamental and its series of overtones, which combine to generate what musicians call the color of a note, its particular timbre and complexity. A cello string generates a series of partial tones by itself: in addition to the wave formed by flexing up and down once along its entire length, it can vibrate twice along its length, sounding the first overtone, an octave above the fundamental; it can bend three times, forming the second overtone, a fifth higher than the second—and so on, following the sequence Pythagoras himself identified (as described in chapter one).

The overtone series for each note struck on a cello gets amplified along with the fundamental, but the behavior of the cello's body determines which overtones get emphasized, and which fade. The cello's various components respond to the motion of the string by bending and bowing interactively, communicating vibrations back and forth. The particular intensity of the different patterns of movement determines how much of the energy put into the system by the stroke of the bow reaches each wavelength, each partial of the tone being played. That distribution of energy from the fundamental up across the entire overtone series creates the actual sound, a complex note with the particular qualities of richness, color and nuance that each instrument displays. Finally—and instantly, following the stroke of the bow—the original motion of the string, now transformed into an oscillating mass of air within the resonating body of the cello, emerges from the two f-shaped holes carved into the instrument's belly. That vibration sweeps outward, filling a room, sounding a note.

If that note emerges from a Stradivari cello, it will have a distinct, identifiable set of characteristics. The myth of Stradivari has it that he possessed some mysterious gift that enabled him to massage his constructions into the realm of the miraculous. But the reality is more prosaic: to put it in the language of physical acoustics, what makes Stradivari instruments special is the particular mix of overtones and fundamentals they produce, along with the capacity of the finest instruments to generate many different such mixes. Accompanying such qualities of tone, Stravidari's cellos also possess an exceptional (though not unique) ability to produce sound that carries, able to fill even very large halls. Stradivari achieved those results by experimenting with each of the major variables that shape the performance of the machine that is a cello. His process (if not its result) holds almost no mysteries: Stradivari was essentially a craftsman of science, one with considerable, demonstrable knowledge of mathematics and acoustical physics. The surviving physical evidence of both the instruments themselves and the tools used to build them, combined with the analysis of the detailed design of his best creations, has made it possible to retrace, to a surprising degree, the sequence of intellectual and technological ideas Stradivari used to create his unique cellos. We know what he did—we have the instruments—and we know almost completely how he did it.

That knowledge has emerged from over a century of research on two fronts: an archaeological approach that tries to recover the step-by-step sequence of Stradivari's craftsmanship, and an analytical attack that seeks to identify the underlying physics that could cause a machine like a cello to produce a sound like Stradivari's. The best description of Stradivari's methods was assembled over a lifetime of investigation by Simone Sacconi, a violin maker, restorer, and copyist. Sacconi was a devotee in a one-man cult whose deity was Stradivari. He committed himself to his life's work when he was eight: at that age he opened up a violin owned by his father, an orchestra violinist, to see how it worked. His father sensibly safeguarded his remaining instruments by enrolling Sacconi in a violin-making apprenticeship. He encountered his first Stradivari when he was thirteen. He measured and drew it and continued to study and copy Stradivari instruments for the next sixty years. Finally, in 1972, he published his conclusions under the mocking title *The "Secrets" of Stradivari*.

As Sacconi reconstructed the process, Stradivari started each project by drawing the outline of the instrument called for on paper, and then built a mold out of wood that could serve as a template for the construction process to follow. Such molds revealed Stradivari's scientific side, at least to Sacconi. He found that Stradivari clearly understood and used quite advanced mathematical ideas, deriving the precise proportions of his molds by working through several geometrical operations. Having fixed the outline of an instrument and built the mold, the next step, as Sacconi reconstructed it, was to build the sides of the instruments, the ribs. Stradivari worked with strips of maple, as thin as one millimeter, using a hot iron and humidity to shape the ribs around the hourglass curves of the instrument.

The next major task was to carve the back and belly into the correct form. Bellies were made of spruce or occasionally fir; backs, like the ribs, were usually made of maple, though some

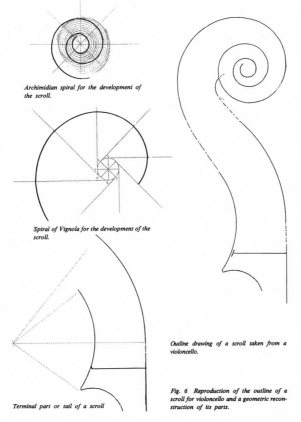

Archimidian spiral for the development of the scroll.

Spiral of Vignola for the development of the scroll.

Terminal part or tail of a scroll

Outline drawing of a scroll taken from a violoncello.

Fig. 6 Reproduction of the outline of a scroll for violoncello and a geometric reconstruction of its parts.

Sacconi's diagrams of Stradivari's scrolls for a violoncello.

cellos used poplar or willow wood. That wood may have been treated, aged, or seasoned—one theory suggests that wood stored in Venetian canals, for example, would have encountered microbes that could alter the acoustical properties of lumber—but so far, no attempt to re-create those conditions has produced an instant Stradivari. According to Sacconi, the desired sound emerged in the building: he argued that one of Stradivari's great insights was to build unusually thin components for his instruments, back, bellies, and ribs that could vibrate freely, rapidly, responsively. Stradivari's cello backs measure about seven and a half millimeters thick at the center, substantially thinner than more recent instruments. The critical decision to be made was the precise arching or curve outward to be used for both back and belly. Cello bellies and backs were made of two, sometimes more, pieces of wood, shaped and joined. Each piece would be cut into a slab of approximately the thickness of the thickest part of the component being made, and then would be carved, first roughly, and then with fine finish work, to the exact thickness and shape required. Stradivari's craft skills here showed themselves at their best: he was able to carve a solid piece of maple into a symmetrical arching with striking precision. That task is complex, as the shape of a cello, with a top portion, a narrow waist, and a bottom slightly wider than the top, means that that curve of the arching has to vary up and down the instrument to retain the desired thicknesses.

Once the components of a cello's resonating chamber were complete, Stradivari next cut the f-shaped holes, whose position he determined by considering the curve of the arching, the volume of the resonating chamber, the strength of the wood being used, and the position in which the bridge would be placed. Next he would add the bass bar to the belly of the instrument, and then prepare to close the sound box, placing the sound post and gluing together the cello. Several steps remained—carving the inlaid outline, called purfling, that appears around the edges of the belly and back, carving the spiral scroll at the head of the neck of the instrument, building the neck and fingerboard, adding the pegs, and so on. But with the closing of the cello's body, Stradivari came to the end of the fundamental decisions that determined the acoustics of the instrument.

In sum, as Sacconi retraced Stradivari's style, the critical features that distinguished him from his contemporaries and his successors were the lightness of his constructions, the particular dimensions of his archings, and the experimentation with the volume of air to be enclosed by the instrument's body. Stradivari's evolving choices in each of these areas suggest not only that he knew geometry but that he understood a great deal of practical acoustics—and was especially sensitive to the relationships between the major variables that shaped the sound of his creations. Sacconi reported that Stradivari was prepared to shift the position of his f-holes each time he altered his belly archings, for example, and more generally, he noted that every change Stradivari made in one of his molds rippled through all the dimensions of the instrument built on the reformed template. Even the decorative scrolls, Sacconi argued, revealed Stradivari's deep, systematic knowledge of his craft. Their form emerges from the intersection of two spirals: Archimedes' spiral shapes the inner coil of the scroll, while Vignola's forms the outline of the large curve around the back of the scroll. The resulting shape looks

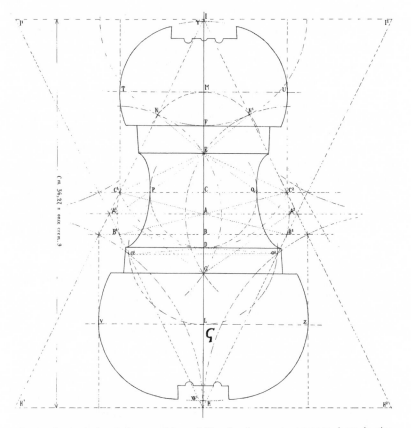

Sacconi's analysis of the precision of Stradivari's measurements, here in the outline for a violin mold.

graceful and effortless; but, wrote Sacconi, one has to know the principle involved: "It cannot be done by eye or freehand."

This is where Sacconi's account opens the door for the analytical approach to Stradivari. Sacconi argued, in effect, that Stradivari had mastered his physics as a practical matter—and thus far the case rings true, for while one great instrument might be an accident, hundreds cannot be, which certainly suggests that Stradivari knew what he was doing. Given that, others who also know physics ought to be able to recapture the principles with which Stradivari worked. The main line of inquiry has been to look at the acoustical behavior of the disassembled components before they have been incorporated into a completed instrument. Instrument builders routinely tap the belly and back plates of cellos, violas, and violins that are under construction, listening to the pitch and clarity of the sound those plates make when struck. In the last twenty years, detailed, quantifiable analysis of both the motion of the tapped plates and of the tones that result has replaced the craftsman's knuckle and ear, and two categories of data have emerged.

First, a variety of imaging techniques has detailed a series of different patterns, or modes that occur on freestanding belly and back plates when they are struck with blows of varying intensity. Second, the pitches that sound have been analyzed to yield the basis of a theory of violin-family construction. The fundamental pitch of a back or belly changes with the shape of the plate. It is thus possible to "tune" a violin-family body by altering the thickness or arching of the plates. Stradivari's "secret," according to this kind of analysis, was the precise tuning he used on each plate to arrive at a perfectly balanced instrument. Several makers have tried to build Stradivari-like instruments by re-creating back and belly tunings observed on various Stradivari violins. More generally, Carleen Hutchins, a violin maker who has pioneered the mathematical analysis of violin acoustics, argues that an instrument builder can achieve the best results by tuning the top plate of a violin-family instrument vibrating in one pattern, the "x" mode, to a pitch one octave above the bottom plate, vibrating in another pattern, the "ring" mode.

Using this kind of method, Hutchins, collaborating with fellow violin researcher Frederick Saunders and others, began to examine the characteristics that would produce the "ideal" instrument—attempting, in effect, to leap past the accumulation of craft technique represented by Stradivari and the Cremonese masters, through the application of the first principles of acoustical physics. The group focused their studies on the scaling of members of the violin family, examining the acoustical relationships between the sizes of the bodies of the different instruments—violin, viola, cello, and their deep cousin, the double bass. The group noted, correctly, that the viola in particular had a body that was considerably smaller, relative to its range of pitches, than that of the violin. The cello, too, turned out to be cramped, given the notes it had to produce, though less so than the viola. To design a new family of violin-style instruments with more consistent resonating chambers, the group turned to one of the simplest rules in the physics of resonating bodies: the general law of similarity. That law states that if a resonator's linear dimensions increase by some number A, then the natural frequencies at which it vibrates decrease by the reciprocal number 1/A. The bigger the box, that is, the lower its fundamental note, the pitch decreasing in inverse proportion to the change in size. The law holds for simple resonators, like a freely vibrating flat plate, and is approximately correct for the more complicated shapes of the violin and its siblings.

With this scaling law in hand, the acoustical errors of traditional instrument design become obvious. A cello, for example, with its bottom note one-third the frequency of a violin's lowest string, ought to be three times as big as a violin in all its linear dimensions. Such an instrument would have to be about five feet, ten inches long, dwarfing a conventional cello, which measures only about four feet in overall length. An instrument that big would be extremely cumbersome to play, and a correctly scaled double bass would be even worse, but Hutchins and her colleagues did develop a family of eight violinlike instruments, covering the entire pitch range of orchestral music. They made a number of compromises to ease the use of their new designs, and their instruments did not scale precisely: the volume of their baritone instrument, tuned to the cello tuning, is less than three times the size of their violin equivalent. But by manipulating length, depth, and width, they did generate instruments that came closer to acoustically optimal dimensions than those of any previous makers, Stradivari included.

Several sets of these violinesque instruments now exist, and they do perform the task for which they were built: they are playable, and they produce sounds whose properties can be clearly related to their design. The violin in the new family, for example, is slightly larger than the traditional version, and it consistently produces louder sound—an average increase of three to four decibels—than does an ordinary violin. Because each of the instruments is scaled along consistent lines, the instruments possess a similarity of tone, which has created an ensemble sound that many audiences have found pleasing. In a recent demonstration of the characteristics of such instruments, Yo Yo Ma performed in Amsterdam using a Hutchins-designed alto violin or "vertical viola" for the Benjamin Britten concerto for viola. The Hutchins instrument actually looks like a viola on steroids— just a little larger in all dimensions than its conventional cousin. It is played cello style, perched on top of a Pinocchio-length end pin, and its tone possesses significantly more power than that of ordinary violas. In Ma's hands, the instrument produced a strik-

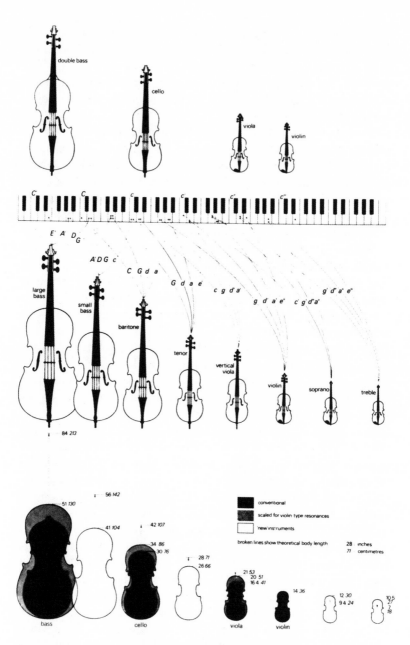

The Hutchins family of instruments alongside the conventional violin family.

ing effect—an almost harsh, strident sound that seemed to leap out into the hall. In the context of Britten's score that edge to the solo part worked beautifully. But what is lost to purchase such punch is the subtlety of expression the great old instruments produced. Stradivari's work, like that of the other old masters, turned on the expressive range of his instruments—their ability to generate a wide variety of sounds, to be used by the skilled player to express a full palette of musical effects and emotional colors. The new instruments provide all the power and clarity one could desire, but compared with the old instruments they are monochromatic and dull.

Politely, this attempt to derive the perfect instrument from the laws of physics—an instance of the top-down approach that seeks to construct particular phenomena from a knowledge of the general rule—has not, in fact, supplanted the old Italian master's work as the ideal. More bluntly, as one distinguished Stradivari scholar and instrument builder put it, this kind of work demonstrates "the nonsense of the 'scientific approach.' " Nonetheless, it still leaves another strategy open. The bottom-up approach to the mystery of Stradivari begins with the particular instruments he built—and the simplest, most obvious first step toward recreating their sound is to copy them as precisely as possible. Beginning around 1800, when Stradivari's instruments were first generally regarded as the best by Cremonese makers (and hence the best available), highly regarded violin makers began to emulate the particular Stradivari instruments that pleased them the most, producing virtually identical physical replicas of the old instruments. The best of these display excellent craftsmanship and fine appearance, but they all fail the fundamental test: they may look like the originals, but they do not sound like them. The Hill brothers examined several of the highest-rated attempts of nineteenth-century imitators and dismissed them all—and their judgment holds, broadly, for more recent copies. The instruments they tested could be very close to the originals by all ordinary criteria: the instruments built by the Guadagnini and the Gagliano family were "constructed on the principles of Stradivari [and] the material used is in many cases acoustically equal." Despite that, noted the Hills, "they have by no means the same character of tone." They shouldered aside the French maker Lupot, saying of his work: "The tone is . . . of a stiff and veiled character." With a commendable lack of nationalist prejudice, they rejected the English builder Daniel Parker's long-strad copies, complaining of the "metallic character" of their sound.

Such criticism does have one flaw, though. All this talk of tone and color via the use of subjective adjectives like "veiled" or "metallic" is part of the mystique of old violins. But this kind of language also works to preserve a sense of esoteric, privileged knowledge that attends any story of ancient traditions and the decayed state of modern ideas, and it can be used to keep out people who have not been admitted to the mysteries. The actual situation is that the difference between modern instruments and old ones is readily accessible. Untrained ears can hear it, and though the language that describes the sound is imprecise, qualitative differences between old instruments and new can be measured and quantified with ease. As machines—not magical implements—old violins behave in distinct ways that separate them from new ones. Studies of the frequency response of three types of violins—mass-produced modern instruments, new instruments made by master builders, and old Italian violins—yield snapshots of the kinds of sound each category produces. The factory-built instruments are thin and weak in the lower register, and their response steadily declines in the higher range. The handmade instruments perform strikingly differently. They have much more bass, and they reach peak response relatively low in their range, around the frequency of 1,000 hertz, or cycles, per second. They then maintain more or less that same sound output all the rest of the way up to their highest notes. The old violins display a much more varied curve. Their bass is marked by three distinct peaks, with valleys in between. They then ramp up to a peak much higher than do newer violins, around 2,500 hertz, before falling off in a sharp dive.

In other words, old violins display much more complicated acoustical behavior than do even well-made modern instruments—which confirms what the Hill brothers' sensitive ears already heard: that the older instruments possess a greater potential for varied expression than do newer ones. Two other bits of data fill in the picture of what the old masters accomplished. The peak response of the older violins falls within the most sensitive range of human hearing, and the location of that peak also bears a striking similarity to a peak region found in the frequency-response curves of trained operatic voices. It is often said, to return to the imprecision of metaphor, that instruments of the violin family "sing." By the evidence of precise acoustical analysis, the best of the old violins, violas, and cellos match the metaphor, and recall that most emotionally expressive of musical instruments, the human voice.

Recognizing the complexity of the sound produced by the older instruments, though, merely defines the problem more sharply than before. The question remains: How did Stradivari and his contemporaries get that sound? The repeated disappointments of even the best copies of Stradivari suggest that the old masters had some critical knowledge, lost to modern builders, that made their instruments more than the sum of their (by now) exhaustively analyzed parts.

The search for that secret has come to rest on the final stage in violin-family construction. After Stradivari worked out the precise design of an instrument and assembled its components, he finished the instrument, protecting the naked wood with a coating of varnish. Stradivari was one of the last makers to use the old-style Cremonese finish, as the composition of instrument varnishes changed abruptly around 1750, soon after his death. No written version of the older recipe remains. Consequently, the Cremona varnish has become a creature of legend, the magic bullet that could transform any instrument (or maker) into a Stradivari. That legend must be at least partly true: the Cremonese varnish does affect the final sound quality of an instrument. Sacconi reported locating a Stradivari violin that had been revarnished with a modern finish, which made the instrument "unrecognizable both in its sound and its aesthetic characteristics." A varnish, any varnish, shapes the sound of an instrument by altering the flexibility of the wood. A thick, heavy finish damps vibrations, muffling the sound; one that is hard and rigid confers a metallic, harsh tone. Stradivari probably achieved the best results of any Italian maker, but in common all the old varnishes managed to reach a beautiful compromise. They conferred strength and rigidity to the wood of the resonating chamber, allowing the brightness of tone to show through, while dampening the vibrations just enough to keep the instrument from screeching and sounding shrill.

Yet the notion that the varnish was the one critical mystery of their trade would have come as news to those who used it. It was hardly treated as a rigorously protected trade secret, as were the techniques of porcelain making that the Chinese attempted to hide from European spies. While each maker might have had his own variation, the basic ingredients were common knowledge and were available on the open market. Records show that varnishes were mixed by assistants, not masters, which suggests that even individual recipes were hardly private. Jacob Stainer, the best German violin maker, never lived in Cremona, yet he possessed the "secret" of the Italian varnishes. We do not know what they used or how they used it, but they knew, and apparently they saw nothing exceptional about the information.

For modern researchers, recognizing that the varnish recipes were common knowledge aided the search for what was apparently too familiar to be written down. Though the precise Stradivari technique has been lost, a number of researchers have isolated at least the basic composition of the old varnish and have reconstructed the methods of its use. The first critical piece of the puzzle was the discovery, made by Sacconi and others, that the Cremonese violin makers used a two-step finishing process, a method lifted from the work of furniture makers in the sixteenth century. (The job was the same for both crafts, after all: a violin, like a chest, is an assembly of wooden parts, on which the finish serves as a barrier between the easily marred raw wood and daily use and abuse.) The undervarnish was a kind of hardening glue that coated and strengthened the wood surface without clogging the microscopic pores of the material—properties that help to account for the "live," readily vibrating action of the resonating surfaces of Stradivari's instruments. Chemical analysis of wood taken from old instruments has revealed unusual amounts of silica and potash, which, when heated and then allowed to harden into a silicate of potassium and calcium, would form a glassy, hard surface.

Sacconi attempted a bit of experimental archaeology of his own to reconstruct this old glue. At his home on Long Island he grew his own grape vines, enabling him to imitate the old Italian process of extracting potash from the ashes of grape pressings or vine cuttings. Melted with silica and potash, dissolved in water and then boiled, the mixture yielded a clear fluid which, when painted onto a wood surface and exposed to air, turned a golden yellow color. Again, imitating old craft methods, Sacconi would smooth the varnish using a piece of fish skin and a powder of goose grass. This gluey underlayer, like the Cremonese original from which it descended, created a hard, impermeable coat over the instrument, providing the ground for the application of the next layer, the colored varnish.

As was the case with the undercoat, the basic composition of this layer is also well known, and has been for at least a century. The Hill brothers noted that "Stradivari used solely a pure oil varnish, the composition of which consisted of a gum soluble in oil, possessing good drying qualities, with the addition of coloring ingredients." The particular gums and oils were derived from naturally occurring substances: propolis, a resinous substance collected by the bees of Lombardy from poplar tree blossoms; turpentine extracted from the resin of larch trees; oil of spike lavender; ethyl alcohol; and others. Coloring agents included cinnabar (reddish-tinted mercuric sulfide, the most common ore of mercury) and red madder (a pigment extracted from the root of the plant *Rubia tinctorum*).

Finishing an instrument required several coats of this varnish, laid on with a brush or a sponge. The wait for each layer to dry could be tedious—in one of only two of his surviving letters, Stradivari explained the delay in delivery of an instrument because of the unusually slow pace of the drying process. The difficulty in applying the varnish at just the right thickness per layer and impatience that grew, waiting for each coat to dry, probably accounted for the disappearance of the older oil-based varnishes. Beginning around 1750, spirit-based varnishes that formed a quick-drying hard shell became available. Vastly more convenient than the cumbersome traditional preparations, the new varnishes caught on, especially in the high-volume violin factories of Milan and Naples, despite the lack of depth and body they imparted to the tone of instruments. By the end of the eighteenth century, the remnant of those trained in the workshops of the best Cremonese makers—Stradivari's sons, his assistant, the last Guarneris—had all died, taking with them, apparently, the knowledge of the older technique.

Both the appearance and the tone of Stradivari's finished instruments certainly were the product, at least partially, of the particular qualities his varnish imparted. But if, as Sacconi's own work demonstrated, that varnish was less a secret formula and more a forgotten and mostly recoverable craft technique, then the fundamental mystery of the uniqueness of Stradivari's success remains. Sacconi himself was able to create a finish that displayed many of the properties identified with the Cremona varnishes. He also knew, as well as any other violin expert of his day, the exact measurements of Stradivari's instruments. He devoted much of his life to repairing and restoring the great instruments of the Cremonese masters—and he spent the rest of his time building his own, striving to recapture the unique, unmistakable sound of his beloved Stradivari. Yet he failed, just like his predecessors in the nineteenth century. He did build credible violins, some of which have captured highly respectable prices at rare-instrument auctions, but their measure can be found in the backhanded praise offered by Andrew Dipper, the instrument maker who wrote the preface to Sacconi's book: "Had Mr. Sacconi had the possiblity of working within the . . . uninterrupted tradition [of Cremonese builders] his achievements may have totally surpassed the achievements of Stradivari." Because Sacconi did not have that chance, in other words, his work did not supplant the old master's—and, presumably, we should not have expected it to do so. And yet, again, why not? What is a cello; what is it in some cellos that eludes the power of scientific description, that escapes our reason and our capacity to create it on demand?

The answer, most simply, is that the cello, any cello, is more than the sum of its parts, a fact which the two chief approaches to the problem of Stradivari ignore. At their logical extreme both assume that there is a single, correct solution. The analytical approach—the top-down inquiry that begins from physical theory—assumes that there is an ideal range of acoustical behavior associated with a type of instrument. That perfect sound, in turn, sets the parameters for the design of the instrument perfectly suited to produce it. The bottom-up approach, combining the archaeological investigation of Stradivari's methods with an exact description of his finished instruments, makes the same mistake in a different way. Where the physicist can imagine a universally perfect cello, the archaeologist choses a particular wonderful instrument—but the outcome is the same: the imitator seeks to build toward the single, right answer, defined by whichever actual Stradivari cello he selects as his target.

Within the context of Newtonian science, such assumptions make sense: one of its fundamental tenets held that the behavior of any mechanical system can be characterized, whether in a general form or for each particular example of such a system. Up to a point, that belief holds true, but from it emerges another idea that is far less secure. Those who sought Stradivari's insight believed that understanding what he did, even how he did it, would lead directly to the ability to replicate; more generally, that the description of a system and its current behavior would produce the understanding needed to predict accurately the behavior of such systems in the future. The reality is far more complex and more limiting. The actual experience of constructing a cello points out the flaw, the almost invisible crack, in the foundation of this Newtonian worldview.

Peter and Wendy Moes are among the leading contemporary American makers of cellos, violins, and violas. Based in Connecticut, they have been building instruments for the better part of two decades. They work as a team, constructing two instruments at a time. They divide some of the tasks. Wendy always carves the scrolls while Peter is responsible for the f-holes and the arching; but mostly they mix and match, doing what needs to be done. It takes them about four months to turn out a pair of cellos. They are slower than many makers, but their pace enables them to control as much as possible the character of the sound their instruments will produce. They have a target clearly in mind—not Stradivari's but their own. Peter describes his ideal cello as one with a lot of bass, a top (A) string that doesn't screech, good power throughout the full range of the instrument, and, most important, a complexity and a fullness to the sound so that it can resonate throughout a concert hall without booming. Wendy's metaphor for the tone quality she seeks is that of a circle. A bad cello will produce a note that traces a round outline. A good one will fill in the shape, shading it with texture and nuance.

The Moeses began as repairers and restorers of antique instruments, and even now they will occasionally do some copy work. One recent project had them building modern parts for an old violin that had had its belly smashed and replaced badly late in the eighteenth century. That instrument had been built by one of the Guarneris, and the Moeses explicitly aimed to re-create the behavior and sound its maker had originally sought. But attempting to imitate the Italian masters seems to them not just a bad idea—it would be boring, Peter argues, to strive for someone else's sound rather than your own—but impossible. Efforts to re-create Stradivari's instruments must fail, according to Peter and Wendy Moes, because of a confusion between what a cello does and what it is. The chief conceptual mistake made in the analysis of fine instruments is to treat the components that make up the instrument as discrete, individual parts—an error to be found in all the textbook descriptions of cello anatomy (like the one given above). Rather, as it comes together, a cello accumulates into a single, functional whole, and the behavior of each of the distinct pieces must converge.

If such parts behaved the same way every time, the problem

would evaporate. Find a design that works, stamp out the sub-assemblies like fenders for Buicks, and bolt the whole lot together, one after another. But from the beginning, in the workshop, before a single piece of wood has been cut to size, the image of the cello as a machine plays false. From conception to completion, the process of constructing an instrument more closely mimics living behavior than the mechanical logic of the assembly line or the inflexible, descriptive rigor of physical analysis. It all begins with the wood, and the builder of an instrument must alter where she alteration finds: Wendy and Peter Moes in effect grow, rather than assemble, each cello, constructing each successive part of the instrument in response to the particular, individual characteristics of the bits already built.

Thus, for the Moes duo, the first and most important decision to be made on each instrument is the selection of the individual pieces of lumber that must be persuaded to make beautiful music together. Their one "secret" lies in their insistence on using old wood. They seek out the finest material board by board, through one of the smallest and most specialized timber markets in the world. The maple they use for the backs and ribs of their instruments is at best middle-aged—cut about one hundred fifty years ago. They get the best they can, seeking boards harvested from houses built in Italy or Switzerland that have matured toward the optimum combination of strength and flexibility. The spruce found in their instrument bellies is even more special. In 1978 they helped take apart a four-hundred-year-old house in the Swiss Alps. Its ceiling boards came from stands of spruce that probably served the master craftsmen in the rich Italian cities to the south. That timber comes from spruces that were harvested correctly, felled in December, when the sap has slowed and the wood is at its peak, hard and ready to work. (The traditional period for cutting spruce was around the solstice, December 21, but that date was more a handy reference than a requirement. The point was to begin after the temperature had been cold enough long enough to guarantee that the target trees had well and truly settled into their winter habits.)

Finding a whole ceiling made of the right kind of wood was a stroke of luck for the Moeses' workshop; the harvest from other houses of similar age ensures their supply. At that, they are more fortunate than Stradivari himself. He bought a particularly excellent and substantial log of maple in the Cremonese lumber mart early in the 1710s. That log lasted him until around 1720, and proved irreplaceable. Stradivari's violins and cellos from that decade boast the most beautiful backs of any of his instruments, with fine grain and the characteristic stripes of polished maple. After he had exhausted that one log he was never able to achieve that peak appearance again.

For Wendy and Peter Moes, the use of the recycled scraps of old Swiss farmhouses shapes each subsequent decision for each instrument. Even though the ceiling timbers available to them for the moment are from the same lot, each piece of timber differs from every other, and they find themselves cobbling together both backs and bellies out of several pieces of wood. So the first problem is to find lengths of spruce and maple that will, when glued together, function effectively as a single sounding board. A cello under construction in their shop at the time of this writing possessed a belly made of five different pieces of spruce, each with a different grain, with differing strengths both along and across those grains, showing three knots, one large, all of which had to work as a unit.

The process of putting together such a belly is one of ebb and flow. The Moes duo begins with a set design, a cello mold that they use, just like Stradivari and all of his rote imitators, to fix the basic dimensions of their instruments. But to build the five-piece belly sitting in their workshop, the couple had to carve each piece to the particular thicknesses (varying from plank to plank and within each segment). The precise choices depended on what seems like an instinct, a feel, but actually is knowledge acquired by long practice that gives them their sense of the acoustical consequences of each decision. The choices change with every instrument. The wood within the large knot on the lower left of the current belly was far harder and more rigid than the unblemished trunk wood around it, which would have set up a dead spot in the vibrations of the belly as a whole. So Peter carved that one small area down to almost paper thinness and backed up the knot with a hidden spruce patch, glued to the inside of the instrument. A particularly wide-grain slat may need to be slightly thicker than the narrow grain strip to which it is joined. If one belly seems a little more flexible, fully assembled, than the other of the two cellos being built at one time, the arching of backs and fronts may have to differ.

Each cello Wendy and Peter Moes build differs from every other to a greater or lesser degree, just as Stradivari's did; and it is here that the metaphor of growth applies. The plan for a cello corresponds to its genetic code; the specifications guarantee that what will emerge will be a cello, not a viol or some hideous giant mutant ukulele. But as the instrument makes its way from plan to completion, it responds to the particular conditions that shape its unique configuration of wood and glue. The instrument to which Peter applied another coat of varnish as he told its story was a member of the species cello by its nature. In its nurturing at his and Wendy's hands, it gained its distinctive qualities as a fine cello, one with its own voice.

Just as it is impossible, peering at an infant in its crib (or for that matter, at a seedling poking through the forest floor), to predict how a child (or a tree) will turn out, so the Moes pair cannot anticipate exactly what they will need to do to produce each of their instruments. They know, along with every master instrument maker before them, that nothing will permit them to replicate a given instrument exactly—not one of Stradivari's, not one of their own. On the most obvious level, the variability of wood makes the difference. Stradivari himself ran out of his best maple and made do with other logs, with a slightly different result. This vexing inconstancy defeats the strategy of precision measurement: the exact figures for the dimensions of a given cello will produce the performance characteristics of that cello only if the imitator can find wood that will behave precisely as the original timber does. This does not occur; no two slices, even from the same log, share identical properties.

And the way each piece of wood, each carefully shaped component, differs from the next also undermines those trying to copy Stradivari using analytical methods—those attempts to isolate the physics of the Stradivari sound in some "secret" acoustical property of one aspect or another of his design. The point of such investigations is to gain the ability to predict how any proposed cello design will behave, allowing the scientific instrument builder to direct his efforts to generate a specific result. Such predictions are made by taking the information available at one time and using the apparatus developed out of Newton's laws of motion to forecast a system's behavior indefinitely into the future. The failure of this approach in this instance points to one of three conclusions: the investigators have been inept; Newton's laws and the rest of mechanical physics are wrong; or there is a property inherent in cellos that renders predictions about their behavior unreliable. The first two possibilities hold out some hope of eventually discovering the "secret" of Stradivari, but it is the third option that is correct: the Moeses' inability to make two cellos the same derives from a fundamental principle in nature and not simply from the imprecision of human craft.

This is so because cellos belong to a class of phenomena, ubiquitous in nature, called nonlinear systems; their behavior cannot be predicted precisely in the conventional sense. Newton's equations do give reliable accounts of such critical processes as the motion of the solar system. We know where Jupiter will be, to a high level of precision, for thousands of years to come. Newton's laws of motion work perfectly to analyze and predict what are called linear systems, where one unit of energy, for example, makes a ball roll at one unit of speed, while twice as much energy produces twice the velocity. (A system would still be linear if each unit of energy produced only half a unit of velocity, or two units—the key is that the amount of change produced by a constant input does not vary. The solar system, incidentally, is nonlinear: the different bodies in it impose a wild variety of gravitational tugs on everything around them, but big objects, like planets, act like linear phenomena, rolling predictably along their orbits for great lengths of time. The question of whether they must always do so increasingly troubled a growing number of thinkers in the nineteenth century, whose story comes in the next section.)

But if forecasting the future of linear, or nearly linear, systems poses little difficulty—just figure out how much energy is going in and crank out the number for how fast an object will travel or where it will be at any time down the line—nonlinear systems do not behave so neatly. Every component of a cello—back, belly, ribs, sound post, varnish, and the rest—reacts differently when agitated by different amounts of energy. Being hit twice as hard does not necessarily make a belly vibrate twice as fast. The relationship between input and output, energy in and sound out, for each component follows its own nonlinear course. What this means is that a small change in the input of energy to a component can trigger a major change in what the component does, how it vibrates, where it distributes that energy across a back plate, and so on. This property, known as the "sensitive dependence on initial conditions," is what makes prediction such a problem.

What makes it worse, so far as the cello goes, is that the instrument's nonlinear elements work together, making a far more complex nonlinear whole. The interactions between the components produce rapid, continuous variation in the transmission of energy among all the moving parts of the instrument. If a little alteration can produce a big difference in sound, then in order to predict what sounds will emerge from a given cello one would need to know exactly what all those nonlinear components will do in any given circumstance in response to any stroke of a bow. That does not on the face of it seem impossible: it is, after all, simply an issue of precision measurement, a task that we can do very well. Find out the material properties of each piece of wood used, measure their vibrations under different stimuli, and then grind out the calculations (which is what Hutchins and her collaborators have done with their new violin family, more or less). But in fact, such measurements do not produce Stradivaris. Some things, it turns out, are unmeasurable. The sensitive dependence on initial conditions that characterizes a cello's nonlinear performance means that even very slight differences between the observed, measured value and the actual one out there in the body of the cello can off throw the results in a direction that cannot be anticipated. Because all measurements are only as exact as the tool used to make them, errors are inevitable, and those errors have, so far, been enough to prevent anyone from imitating Stradivari.

In current jargon, this falls into the area known as chaos. The term is misleading, for the physicists' chaos possesses a great deal of order and recoverable information, quite at odds with the chaos of Genesis, random, formless, and unmapped. The experience of instrument makers suggests the kind of knowledge accessible in this realm of scientific chaos. Even though Peter and Wendy Moes cannot anticipate exactly what shape the cello they are about to build will take, nor precisely how it will sound, they still understand a great deal. They do know broadly how the kinds of cellos they make will perform. And with the intuition they have honed through years of experience they know how to blend the mixture of all their physically complicated components to produce a cello that will fall into the range of performance they desire. Theirs is the practical version of the kind of prediction or anticipation that the mathematics of nonlinear systems makes possible. Instead of the exact predictions available for linear problems, one can analyze the types of events that can occur within a given nonlinear system—what kinds of cellos can be built, and what factors can constrain the range of options—to narrow in on a particularly Moes-sounding cello.

226

The existence of nonlinearity in a system produces a tension, a kind of balance. All of Newton's laws still govern the system. Its behavior is thus intelligible, restricted, predictable up to a point: a cello is never going to sound like a ukulele; a Moes cello will not mimic one of Stradivari's. Against this knowledge, the sensitivity of many nonlinear systems to the tiniest gusts and buffets produces an unpredictability that can be seen as flexibility, a sense of possibility. The nonlinearity of its response is what distinguishes the sound of a good cello, producing what Wendy Moes imagined as a filled-in circle of sound, rich and expressive. Stradivari's instruments possess this quality of complexity, of personality in the highest degree. And at the same time, instruments built in the course of a seventy-year career still sound recognizably similar, made by the same hand, which means that, as a practical matter, even if he did not do the math, Stradivari understood how to build toward the kind of results he sought. But the elusiveness of those results suggests that Stradivari's cellos produce their music out of the extraordinarily complex interplay of all the nonlinear elements from which the finished instruments were composed. The "secret" of his sound lies not with any one feature but with his ability to marry the lot of them together with a subtlety and precision that cannot be isolated. Someone could yet re-create Stradivari's effects, given time, trial, craftsmanship, and luck. But the inherent unpredictability that permeates the task of building a first-class cello means that the hunt for the law, for the list of rules for building a Stradivari, cannot succeed.

And as for the cello, so for most of nature. Formal scientific recognition of the ubiquity and implications of nonlinearity lagged far behind the unsystematic understanding of instrument builders, but in 1963 a highly theoretical paper appeared in an obscure meteorological journal and pointed out that winds interact with ferocious complexity. Using an extremely simplified computer simulation of the atmosphere, the paper's author, MIT's Edward Lorenz, then tested what would happen if there was a difference between measured weather data and events in his imitation atmosphere. He found that even the slightest of errors would be multiplied by the nonlinear features of the equations that describe motion in the atmosphere, to the point where any weather predictions made with that data would hopelessly diverge from the course the atmosphere actually took. As Lorenz put it, an uncounted flap of butterfly wings in Brazil could trigger an unanticipated typhoon in the South China Sea two weeks later. The same reasoning applies wherever one looks. Ocean currents, the distribution of galaxies, the path of a pinball, wheat yields in the Dakotas, rates of forest decline—all are described by nonlinear formulae, and all evade certain prediction.

Just as in the case of the cello, the lack of certainty built into the nonlinear world of everyday experience does not mean that nothing is understood, nothing can be learned. Rather, the recognition that perfect knowledge of our material environment cannot be attained has forced twentieth century scientists to come to grips once again with the question of what can be known, what science can actually say about nature. That question still lacks a full answer, but some critical pieces of the puzzle have found their places. Along with its pervasive imprecision, modern science has developed the most comprehensive, most complex portrait of the natural world ever developed; science now produces more statements about more phenomena than ever before. To do so, though, it has undergone a shift in method and focus that Newton and his successors would have found incomprehensible. To Newtonians, each question had its singular answer, one that would remain the same no matter who asked it, or why. But now, the uncertainty that undercuts every measurement of some fact in the real world compels the observer to choose which question to ask, which aspect of a phenomenon to study.

The necessity of choice became overwhelmingly apparent when Heisenberg elevated uncertainty to a principle in quantum mechanics in 1927, having recognized that on the subatomic level the observer had to emphasize only one of a pair of properties to study at any one time. In one of the prominent interpretations of quantum mechanics, the idea took on a larger meaning: that in choosing what to study, the scientist in effect creates the object of his inquiry. To Newton, and to every scientist after him down to Albert Einstein, and to most people still, nature exists out there, real, distinct. The notion that one could create something by looking for it—and that one person's view would differ from another's—was a scandal, unacceptable. Yet what Heisenberg uncovered in the realm of the very small recurs, for different reasons, at the level of everyday experience, the level in which the weather blusters and cellos play. The impossibility of constructing a complete, accurate quantitative description of a complex system forces observers to pick which aspects of the system they most wish to understand: to study storminess across a huge system, perhaps, or the course of a single storm; to learn all one can about the back plate of the Duport cello Stradivari built in 1711; or to analyze the universe of sound that cellos can produce.

What one studies from among this wealth of choice depends on what one wants to know; the questions create—or at least determine—the range of possible answers. No such answer can be completely "true": instead of saying, "This is what nature is like," they can claim only, "This is what nature seems like from here"—a vastly diminished claim from that of Newton. The critical issue raised by such subjectivity is how to decide what value each partial answer has, what connection it actually makes between the real world and our understanding of it. The object of study, the focus of much of modern science, has therefore shifted inward, to examine not nature itself but rather to study the abstract representations of nature, the choices made of what to leave in and what to drop out of any given study.

The use of those abstractions, simplified models of systems too complex to be understood by direct observation, represents a new method of doing science, one created in the twentieth century in response to the collapse of the Newtonian dream of certainty. As Newton's revolution replaced the question Why? with the narrower, more tractable What?, so this method shifts the focus again—asking not what exists in nature but how closely our images of nature resemble what we can see out there. What this new approach actually tells us about our surroundings will be taken up in the following chapters. The answer is still evolving, but there is a parable to be drawn from the history of the cello, of Stradivari and of Bach, that suggests how one might begin thinking about it.

Between the years 1717 and 1723, Bach enjoyed a kind of holiday, serving as Kapellmeister, or music master, in the court of Prince Leopold, ruler of the tiny German principality of Anhalt-Cöthen. In a peculiar way it was Bach's good fortune that Anhalt-Cöthen prayed in Calvinist churches, a sect that took its religion straight, with far less musical accompaniment than many denominations. There was thus little demand for sacred music—and the principality had few professionally trained singers able to perform at the standards required. Bach himself was a devout Lutheran; he composed his cantatas, his Passions, and the rest as a joyful gift to God. But in Leopold's capital city of Cöthen, he enjoyed a respite from the incredible labor of feeding music week after week into the church calendar. Thus liberated, Bach produced one of his most sustained outpourings of secular, instrumental compositions, an astounding body of work that includes some of the monuments of his career. The Brandenburg Concertos come from this period, as do some of his greatest keyboard works (excepting his organ music). Bach wrote for the violin, flute, and viol da gamba; he produced compositions for two instruments, for trios responding to the appetite of a sovereign eager to listen and to play. But of all the music he made in those six years, among the most remarkable was the handful of pieces he wrote for an instrument just beginning to make its presence felt, the six suites for solo cello, composed in 1720.

The cello suites are unique. Solo cello music existed before Bach, but as a body of pieces, taken together, nothing like them had been written before him; nothing like them has been written since. There was a gamba player in Leopold's orchestra who doubled as a cellist, so the works were probably performed in Cöthen around the time they were written. But they mostly dropped out of sight after that. They were used as teaching pieces, and cellists would play a single movement, or perhaps as much as one of the suites during a recital, but Pablo Casals was the first since Bach's lifetime to have treated the six pieces as a unified body of work, performing the whole group as part of his concert repertoire. With Casals's enthusiastic proselytizing, the suites began to represent to cellists what the role of Hamlet does for actors: pieces they had to master to lay claim to the first rank of their profession.

Of contemporary soloists, none has a deeper involvement with the suites than Yo Yo Ma. He performed one of them in public when he was seven years old, and he has returned to them again and again. In a landmark performance in 1992, Ma played all six—more than two hours of music—in a single concert at Carnegie Hall and has repeated the feat since. (The Dutch cellist Anner Bylsma is the other contemporary soloist who has repeatedly performed all six suites together.) Ma at Carnegie Hall used his Stradivari, the Davidov, which met the need: the sound from that cello filled the enormous cavern of Carnegie Hall, carrying even when Ma sought to whisper the notes off its strings, conveying every shade of tone and color he imparted to the music. The cumulative effect of the six suites, layered one atop the next, was as if Ma laid out a map of emotion and ideas, and then explored it in ever greater detail, until his audience could recog-

nize the full, rich, intricate range of feeling that Bach had imparted to his score.

For Ma, the suites (and music in general) "are about the primary things we have to deal with, what we know and the terror
and mystery of what we don't: the terror and mystery of death."
Each of the suites conveys a particular story to him. The fifth, his
favorite, tells of "resignation, loss—with a certain hope as well—
but definitely loss." Expressing what he finds in the score Bach
left him requires a balancing act. On one hand, Ma pays extremely close attention to the elaborate structure of the pieces:
"You have to have an organized mind," he says. "If you go for
beautiful sound and gorgeous melodies but ignore the rhythmic
and harmonic pillars; if you are not constantly listening to the
bass line; if you're not aware of the dance beat; if you don't
recognize the multiple patterns that sometimes go with and
sometimes interfere with strong beats—then you can't extract
the joy from the music. You don't get the feeling of large structure; you can't do it, it just doesn't happen." But on the other
hand, too precise a reading and the joy gets lost again. "Perfection means you hit every note, no scratches, playing totally reli-
ably. But that squeezes the life out; it destroys a piece of music.
The element of risk isn't there. In performance you shape, mold,
and sculpt. You follow whatever you can sense of the motion, the
wave, the physicality of a smooth stream of notes. You have to
build the piece with your fingers, mind, instrument. It has to
breathe. Perfection to me is static."

It is here that the experience of the music touches the issues
with which modern science has been grappling for a hundred
years and more. What Bach wrote is fixed, determined, unchanging—and untouchable by the audience. But then Ma, reading the
score and transforming that text into a river of sound, stood between us and Bach. Instead of Bach's voice, we heard Ma's translation of a text into music, into an experience available to a
listener. And each member of the audience in some sense chose
a particular version of the suites heard that night, encountering
the music Ma made from individual, distinct vantage points. The
sense Ma's remarkable performance gave of revealing all there
was to be found in Bach's writing was actually a persuasive illusion; we heard what Ma found that night with ears tuned by our
own concerns. The loss Ma felt in Suite No. Five seemed to this
listener more like resignation, mixed with anger. Either view will
do, both can be located within what Bach composed.

For Bach's suites, read nature—the unchanging raw material of experience, the world out there, inaccessible to direct encounters. Ma and his cello become the instruments that create one image, a reflection of what can be glimpsed on a page of the book of nature. And finally, we—anyone who encounters each rendition of the music—find ourselves in the position of a researcher, an investigator, a scientist late in the twentieth century, studying each such image to try and locate ourselves within that created world of sound.

At any given performance we cannot locate all of Bach, find all the meaning, sound, sense, and feeling that may be contained within that score. We can always return and hear more, hear further into what we already have heard once—but we will never hear it all. That is our loss, inevitably. But with each foray Ma makes into what Bach wrote, each of his readings that we hear and interpret, we gain some new hint of what might exist within that immutable text. Compared with the hope of perfect knowledge our predecessors nourished, the uncertainty that marks the twentieth century seems a retreat. Yet one more encounter with the Bach cello suites points toward the reward that compensates for that defeat. Perfection is static, as Ma says, dull—in music and in science as well. In place of perfection we find "the terror and mystery" of what we do not know, of possibility, and risk.

INTERLUDE:

WE CANNOT

GO BACK TO THAT

At some level, the history of science is the repeated realization that we have asked the wrong questions, that we cannot find out what (we think) we need to know. Scientific revolutionaries gave up on understanding why things happen—the goal of classical and medieval inquiries—to learn what occurs in nature. Yet even that more limited question turned out to be unanswerable in any complete sense.

Such an end to certainty carries with it both pain and boon. Despite the recognition that some questions cannot be answered fully, science in the twentieth century has become vastly more powerful than it has ever been before. We are now that much more able to determine what can and cannot be known about the material world. But against such triumphs lies the abandonment of one of the goals of scientific investigation that has been present since the beginning. For Pythagoras, as for many since, the mathematicization of nature seemed to suggest a path that could lead toward the mind of God—the architect of the universe who seemed to understand human mathematics so well. Once the scientific revolution took hold, the godhead itself faded into the background—but not the ambition of the scientist. For all the piety of many scientific revolutionaries, one of the main impulses of the new science was to encroach on territory that had once belonged exclusively to heaven. Laplace spoke boldly but said what many thought when he boasted that a science that claimed for itself the divine attribute of omniscience had "no need of that hypothesis." As the poem read: "God said, *Let Newton be!*" and spoke no more.

The case study of the cello illustrates the failure of the effort,

the Babel-like collapse of what had seemed the unshakable ed-
ifice of Newtonian determinism. The cello, of course, is simply a
symptom of the deeper pathology, one most clearly diagnosed by
the French mathematician Henri Poincaré. Poincaré, born in
1854, was the last man even to come close to mastering the whole
body of mathematics. He was a master in the art of discovering
general solutions, methods of great abstraction, that could sub-
sume and replace whole tracts of dizzyingly complex mathematics. In the late 1880s, such a thicket presented itself to him in the
form of one of the admitted scandals of Newtonian physics. Newton's laws were supposed to explain the observed behavior of the
solar system, but Newton himself had solved his equations for
the case in which just two bodies—the earth and the moon, for
example—were involved. Newton then made the assumption
that his equations could be solved to predict the paths of more
objects, three, four, the whole shooting match—but neither he
nor anyone else since had proved that such solutions exist. That
meant that no one knew for sure that the solar system was stable,
that no alignment of its various members could cause the planets
to fly off their usual orbits toward who knows where. The search
for the solution Newton did not find became known as the three-
body problem, and it remained an open question for two centuries. Finally, in 1887, King Oscar II of Sweden offered a prize for
any definitive proof of the stability of the solar system.

Poincaré captured the prize—2,500 kronor and a gold medal—
within two years. Before tackling the solar system as a whole, he
first set out to see if he could demonstrate the stability of a
simple three-body system. He never made it to the larger project.
He found instead that it is impossible to predict mathematically
the complete future behavior of a system that contained even as
few as three objects. The effort foundered against two obstacles.
First, the equations of motion that describe the system relate
position, velocity, and time for each object in the system. However, when there are three bodies to be accounted for, the behavior of each object affects the other two all the time, and their
motion in turn affects what happens to the first—so that it becomes impossible to calculate directly, for each moment in time,
what is going on with all three. What can be done is to approximate the answer, to come up with a number that is close but not
uniquely correct. The mathematical technique that can create
such approximations treats a single value (for position, for example) as an infinite series of numbers. Such series can converge or

diverge. If they converge—like the series $1+1/2+1/4+1/8 \ldots$, which adds up to the number 2 after an infinite number of terms—then it may be possible to come up with one right answer to a calculation. Divergent series, though, keep on going till they add up to infinity—and unfortunately the various series that can be used to approximate solutions for the three-body problem are divergent ones. That divergence still permits quite accurate predictions of the behavior of objects in the real world up to some point in time, sooner or later depending on the particular circumstances. We have a strong sense that the sun will rise tomorrow, certainly, and NASA was able to guide the Apollo missions to the moon with great precision. But at some moment the balance tips and the numbers start adding up higher and higher. Once infinities start appearing in steps of calculations, the equations being worked through are said to blow up, ceasing to produce a finite number that corresponds to some physical property like velocity or position. In practical terms, when that happens such equations can no longer provide usable descriptions of events out there in the real world.

Which meant, once he had demonstrated that such numerical monsters lurked within Newton's equations, that Poincaré had won his prize—not for proving the solar system either stable or unstable but for demonstrating that it would be impossible to determine a definitive result. Some years later the Finnish mathematician Karl Frithiof Sundman proved that the three-body problem does have a certain kind of solution. But his was a mathematical rather than a physically significant result, for it neither permits a numerical calculation of the behavior of the system nor one that predicts in detail the actual motion of the three bodies. Poincaré's insight, by contrast, was to transform empirical experience—the continued errors in the prediction of astronomical events, or, from beyond his realm, the inability to build a cello that could produce the precise sounds being sought—and locate the difficulty. His discovery that even the simplest of systems (far simpler than a cello) could generate complex, unpredictable behavior provided an unequivocal proof that errors in the scientific account of nature were built into the fabric of the mathematics that had once been thought to contain all there is to know about the world.

Which result is the founding realization of modern science, distinct from Newton's view. The striking discovery Pythagoras made—that the universe is mathematical, that its observable phenomena can be described by abstract expressions of mathematical relationships—remains true (though mysterious: Why should the universe obey such rules?). But Poincaré demonstrated first what has become a commonplace since, that there is a gap between the abstraction of a scientific analysis and whatever it is that is actually out there—a fissure that is inherent, built into the fabric of the cosmos and into the language of mathematics used to express the abstract relationships that seem to exist within that collection of perceptions, measurements, and data that we call nature.

Poincaré clearly understood the implications of his conclusions. In his book *The Value of Science* he asked: "Does science teach us the true nature of things?" Absolutely not: "Not only science can not teach us the nature of things, but nothing is capable of teaching it to us, and if any god knew it, he could not find the words to express it. Not only can we not divine the response, but if it were given to us we could understand nothing of it." So much for omniscience. Instead, Poincaré argued: "We can not know all facts, and it is necessary to choose those which are worthy of being known."

Worthy? How chosen? By whom? There is an air of anarchy here, a sense in which Poincaré seems to assert that the impossibility of knowing everything means that one must simply wander within the infinite field of knowledge, picking the facts that appeal to some sense one may have that they are "worthy." But Poincaré also noted that it is "necessary" to choose, that in nature not all facts are equal, and that some will lead to a greater, more powerful interpretation of the patterns human beings make of the experiences they have. It is that dynamic tension between necessity and choice that has driven the invention of contemporary science, its methods, and its extraordinary results.

Poincaré died in 1912, long before the implications of his ideas fully sank in. But the loss of certainty that he uncovered was not unique to science, and in his lifetime others began to grapple with the competing claims of freedom and constraint. One consequence of the scientific revolution had been to rip asunder the two dominant methods of representing experience, science, and art. Newtonianism created an image of science as a machine, spitting out knowledge automatically—exactly the picture that evoked the disgust of artists, unable either to deny or to accept the power of the scientific worldview. Walt Whitman, brought face to face with the harsh certainties of celestial mechanics before Poincaré got his hands on the subject, caught perfectly the artist's mixture of resignation and defiance:

> *When I heard the learn'd astronomer,*
> *When the proofs, the figures, were ranged in columns before me,*
> *When I was shown the charts and diagrams, to add, divide,*
>   *and measure them,*
> *When I sitting heard the astronomer where he lectured with*
>   *much applause in the lecture-room,*
> *How soon unaccountable I became tired and sick,*
> *Till rising and gliding out I wander'd off by myself,*
> *In the mystical moist night-air, and from time to time,*
> *Look'd up in perfect silence at the stars.*

But Poincaré, in proving that the charts and diagrams, the figures, columns, and proofs that so wearied Whitman could not contain all in all, restored what the writer had sensed was lost, a connection between the scientist's and the poet's way of knowing the world—a link born of the necessity of choice. At the conclusion of *The Value of Science*, Poincaré wrote: "It is only through science and art that civilization is of value." Science and art together—the juxtaposition is no accident. Art grapples constantly with the problem of freedom—how to construct a single image, a single statement, a song, out of the infinite possibilities of colors, sound, words, form, to fill the page, the silence. The construction of an aesthetic order out of the entire realm of the imagination presents the same challenge as that of finding the "worthy" facts, and assembling them into a meaningful pattern. Figuring out the methods to accomplish such work is not obvious—it has taken the better part of this century for most of science to come to grips with the problem. Poincaré himself did little more than hint around the issue, but a sense of how it might be done comes from the experience of one of those artists whose confrontation with the warring impulses of choice and fate scandalized Paris just as Henri Poincaré's life came to an end.

On May 29, 1913, Sergei Diaghilev's Ballet Russe presented the premiere of a new work, choreographed by Nijinsky, to a score written by Igor Stravinsky, titled *The Rite of Spring*. In legend, the premiere incited a riot, but the reports appear to have been exaggerated. There may have been some catcalls, but apparently no storming of the stage. But it is clear that the audience was unprepared for the radical new music presented to them by a composer addressing what he saw as a fundamental crisis in the understanding of musical expression, in what music contains and can say to its audience.

For his part, Stravinsky clearly knew what he was after. In an interview he gave later that year, he said that the new piece, unlike his earlier ballets, "no longer calls to mind fairy tales, nor human misery or joy." Instead, the composer had consciously pushed himself along the same track that Poincaré had followed in his mathematical work—"toward a little more vast abstraction." Stravinsky accepted with equanimity his audience's reaction. Just after the premiere he wrote to a friend: "We must wait for a long time before the public becomes accustomed to our language—of the value of what we have done I am certain."

There are obvious reasons why the public did not take to *The Rite of Spring* at first. The piece contained striking harmonies, wild dissonance, unfamiliar orchestration. When Leonard Bernstein analyzed the rhythms in just one measure of the piece, he found demoniac complications that matched, for the first time in five hundred years, the rhythmic intricacies of late medieval music. On top of a fundamental rhythm of six beats, Stravinsky superimposed two more sets of rhythms, each running against the other. One set included groups of four and eight beats only, running over the fundamental six. The other set was far more complicated. The timpani struck out twelve to the bar—but in four groups of three, denoted by a sharp accent on the first beat of each triplet. That set up a four against six rhythm—which another percussion instrument doubled, pounding eight beats to the bar—eight against six, eight against twelve. This would be the voice with which madness speaks, but for the rigor, the control, Stravinsky exercised over his music.

And it sounds great, now, with the hindsight (hindhearing) of the century and all manner of musical innovation to give it context; the public has developed an ear for Stravinsky's language. But Stravinsky sought to express something in that language, something that could not be stated without recourse to terms unfamiliar to his listeners. His *Rite* seemed to them random, a cacophony, meaningless—and yet Stravinsky was confident that it was ultimately intelligible: that in time his audience would be able to recognize in the work a genuine, meaningful musical statement.

Stravinsky's belief rested on his fundamental conception of how music must be made. Before a composition takes shape, there is a moment, an illusion, when anything is possible. It is, however, a trap. Stravinsky, in his Norton Lectures at Harvard University (published under the title *Poetics of Music*) offered what he termed his musical confessions. He admitted to experiencing "a sort of terror when, at the moment of setting to work and finding myself before the infinitude of possibilities that present themselves, I have the feeling that everything is permissible to me." That way lies disaster. "If everything is permissible ... the best and the worst; if nothing offers me any resistance, then any effort is inconceivable ... and consequently every undertaking becomes futile."

There is a rescue at hand, though: "I shall not succumb. I ... shall be reassured by the thought that I have the seven notes of the scale and its chromatic intervals at my disposal, that strong and weak accents are within my reach, and that in all of these I possess solid and concrete elements which offer me a field of experiences just as vast as the upsetting and dizzy infinitude that had just frightened me." With that, the composer's course is clear. "It is into this field that I shall sink my roots, fully convinced that combinations which have at their disposal twelve sounds in each octave and all possible rhythmic varieties promise me riches that all the activity of human genius will never exhaust."

Freedom—the riches that genius cannot exhaust—exercised freely among the concrete elements at his disposal: a strong accent, for example, that will produce one sound juxtaposed with another strong accent, and a different one, no choice about the matter, when played against a weak accent. As Stravinsky put it, when his material imposes its limitations on him, "I must in turn impose mine upon it," which meant, "Here we are, whether we like it or not, in the realm of necessity." And this interplay between freedom and necessity had to take place in order to produce the music. "In art as in everything else, one can build only upon a resisting foundation ... my freedom thus consists in my moving about within the narrow frame that I have assigned myself for each one of my undertakings."

The scientist in Stravinsky sought rigor, precision, the exactitude of expression that the interplay of constraint with his imagination permitted. The art within contemporary science follows the same course. Not all choices are possible, but human genius (we now know) cannot exhaust the choices that exist. Each attempt to make sense of our vast field of experiences resides within a narrow frame, presents just one small echo of the whole. And as Stravinsky knew (as did Bach before him), this freedom of choice exercised against the resisting foundation of nature holds out the chance to experience what we never knew existed—to hear for the first time (to riot or not, as the occasion demands) some new expression of what Poincaré (evoking Newton and Pythagoras, too) recognized as the "universal harmony" of nature.

Thus we compensate ourselves for the loss of the hope of certain knowledge. It is not always enough, of course. We still want to *know*, to possess absolute confidence that the earth will spin around the sun on its stable path forever, to peer into the future as far and as clearly as we might desire. But we cannot, for all that the new language of science to which we have accommodated ourselves across this long century has enabled us to create an extraordinarily rich portrait of nature.

Wallace Stevens, in a magnificent poem called "Connoisseur of Chaos," anticipated this whole drama of loss, gain, and longing. The old ways cannot work: "After all the pretty contrast of life and death/Proves that these opposite things partake of one,/At least that was the theory, when Bishops' books/Resolved the world. We cannot go back to that." Instead there is the present reality, grim though it may be: "The squirming facts exceed the squamous mind,/If one may say so." But Stevens saw his way out of the trap, the endless sea of excess facts: "And yet relation appears,/A small relation expanding like the shade/Of a cloud on sand, a shape on the side of a hill." The trick is to adopt (or better, to possess) the right frame of mind: "The pensive man. . . He sees that eagle float/For which the intricate Alps are a single nest." The obligation, in the world as in the poem, is to look—and to think.

P A R T   3

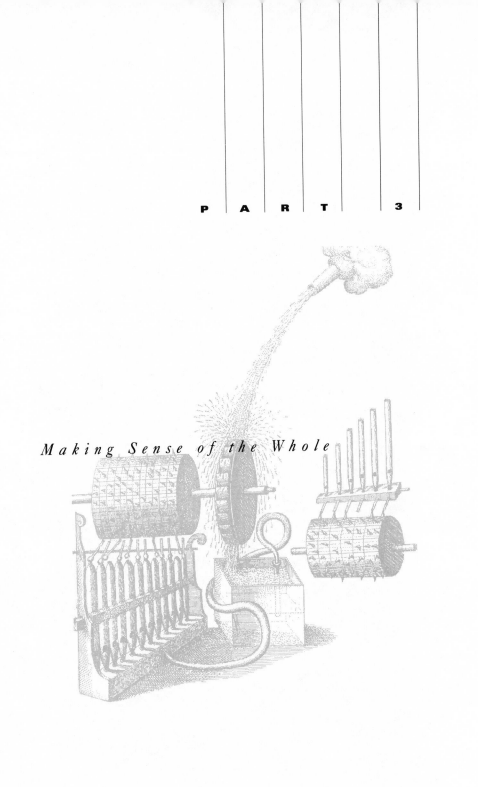

Making Sense of the Whole

Chapter 7

LIGHT

IN THE

CAVE

The laboratory stands on the eastern edge of Palo Alto, on landfill at the margin of the southern limb of the San Francisco Bay. It is a windowless cave, a chamber within a chamber. The innermost isolation unit holds the apparatus of the experiment. The investigators work in full protective gear: gowns, hoods, double gloves. The air they breathe is filtered and cycled so that none may escape untreated into the rooms beyond the laboratory, nor to the atmosphere at large. Passage out takes the researchers through an airlock—a series of four doors with different pressure gradients, past the coded locks and into the familiar world of hallways and fluorescent lights, windows, and sun, and the onshore breezes that blow unaided by any machine.

At the heart of these layers upon layers of protection stands an ancient myth, made flesh by contemporary science. It is a chimera, which was to the Greeks a fire-breathing monster assembled out of the head of a lion, the body of a goat, and the tail of a snake. In its modern form, a chimera possesses tissues from two or more distinct species; the particular version hidden inside its custom-built cell by the bay is a mouse, to which a research team has grafted the components of a human immune system. Such mice can be—have been—infected with the human immunodeficiency virus type one, HIV-1, the agent believed to cause AIDS. The maximum security conditions within which these chimerae live exist to keep that monster securely within its cage.

This organism, thus constructed, has a name: it is the SCID-hu mouse, for Severe Combined ImmunoDefeciency and hu-man. The mice look like ordinary lab specimens—small, white, with pinkish ears. They derive from a colony of specially preserved

A medieval illumination of a griffin—a chimera-like animal with the head and wings of an eagle and the body of a lion.

mutant animals whose genetic defect has rendered inoperable their immune systems—the body's defenses against disease. As simple SCID mice, before the admixture of any human components, the animals are doomed by their vulnerability to infection. They survive past the first few months of life only if they receive a steady diet of antibiotics. Given the correct human tissues, the mice regain the ability to respond to disease that was lost in the original mutation. In so doing (or rather being, for it is what they are that matters) SCID-hu mice have become a powerful new tool for the investigation of immune function and failure. The mice form a scientific instrument, one used, among other purposes, to pursue the secrets of HIV infection.

An instrument, yes—but a mouse is a strange one, surely. Traditional instruments—the microscope, for example, or the more modern centrifuge—extend human senses artificially. The lens sees farther, captures more detail; the spinning chamber of a centrifuge exerts a force beyond the capacity of human muscle power and can separate disparate substances far more delicately than would any human touch. Yet the tasks performed by such instruments are clearly the same sort of thing we do when we look, stir, weigh, count, touch. Not so the mouse. Scientists do things to it; events occur within it; consequences emerge and can be scrutinized. The facts that emerge are different from what we ordinarily imagine scientific facts to be. What the mouse can reveal comes not from nature itself but from a system engineered to resemble the natural phenomenon the mouse's inventors actually want to study. The mouse forms, that is, a model, a kind of metainstrument, different from conventional instruments precisely because it becomes the object, not the tool, of the investigation. The SCID-hu chimera forms a world of its own—a simplified, abstract construct of the real world—within which data is made, not found.

In that act of making its picture of events in nature, the mouse model represents a general change in how contemporary science performs its tasks, compared to earlier methods. The history of the SCID-hu suggests at least part of the reason contemporary science has produced such a rapid increase in the scope and reach of its investigations: the mouse, like all analogous scientific models, explores a set of problems too complex for direct analysis. But at the same time the story of the mouse contains within it an account of where the limits to our powers of inquiry fall, what such models omit from our understanding of the world those models have been built to embody.

Models themselves, of course, have been around since the origin of science itself. At the very beginning of serious speculation on the cosmos, Thales of Miletus produced a model that anticipated, distantly, the role modern models play. He proposed that the universe was made of water, and that all matter, like the earth, could be understood as eddies or turbulence in the flow of the cosmic river. The metaphor—the universe as a whole behaves like the sea that we can see—is a kind of model. The image of a universe of water helped Thales to isolate the critical features any account of the universe would have to include: the natural world's twin faces of constancy and change.

Later thinkers sought to achieve a more realistic effect, to do more with their models than simply reason by analogy. The Pythagoreans transformed the musical scale into a cosmological model. Such a model does more than suggest a resemblance—it makes a claim to explain how the universe works. The idea of the harmony of the spheres provided a schematic account of the shape of nature, within which a reasoning mind could examine the relationships connecting phenomena governed by the operation of its musical law.

A model in this early sense suggested ways to think about what might happen in a system that follows a given set of rules. If the universe follows harmonic laws, then some pattern of planetary movement must make music—and all that remains is for someone to find the tune. The conclusions depend on one's assumptions, of course, but this is the basic idea behind the search for laws of nature. The laws themselves form a kind of model of material phenomena. Thus when Kepler, at his long remove from Pythagoras, successfully (as he saw it) identified the polyphonic choruses of the solar system using his laws of planetary motion, that discovery made sense to him. As he reminded his audience, it was only fit that God employ the most advanced musical technique human invention had yet devised: earthly music ought to resemble that of heaven.

From one end to the other, for Kepler as for Pythagoras, the goal remained basically the same: to find a coherent model of the universe, a consistent, complete plan that would reveal two types of knowledge—practical results, like a simplified apparatus for predicting planetary positions, and aesthetic ones, which would reveal the purpose of the arrangement of the universe, a transcendent order or the mind of God.

In this context the Newtonian achievement was much more than simply a revolution in method. Along with all of his particular discoveries Newton invented a novel form of model, one that still dominates much of the way we think in all spheres. As usual, the new contains within it a kernel of the old. Newton the revolutionary eliminated almost (but, crucially, not quite) all of the analogies his predecessors had used. The Newtonian answer to the twin questions of what the world is and how it works took the form of differential equations, mathematical expressions that relate the rate of change of one process to that of others, locating in the relationships the chain of cause and effect that bound the whole lot, the whole universe, together.

Such equations create what modern scientists have usually meant by the term *model:* a mathematical system whose components represent natural phenomena, and whose results, the solutions to the equations, correspond to, and, in fact, predict, the behavior of those phenomena. Newton's laws of motion and of gravitation thus form a model of the solar system, as had the arithmetic of the scale for Pythagoras. The difference lies, though, with what the two models can do, what they produce: where Pythagoras's version was static, a fixed image of a universe bound to the notes of the scale, the Newtonian construct was dynamic, containing a description of change, a means of plotting the future course of events within the solar system in terms of the changeless formulae of the mathematics of motion.

One of the persistent arguments in the history of thought is over what such formulae, such equations, contain. They can be seen as pure abstraction, a system of symbols whose behavior simply coincides with what can be observed. But dedicated Newtonians held the view that such mathematics are part of nature— that, as Henri Poincaré noted, it was possible at the extreme to believe that the universe itself "is a differential equation." Such a model of nature banishes the taint of analogy: the motion of the planets does not follow a course that is *like* the trajectory mapped by Newton's equations. The two paths are the same. The mathematics, like the planets themselves, were to be observed in nature, extracted from it, understood, and reconstituted to form an understanding of how nature constituted itself.

The Newtonian triumph, then, is this discovery of nature itself, a new vision of nature, a richer and deeper reality, in the form of the mathematical model that could be shown to contain nature's phenomena within its compact and elegant language. Newtonian models seemed to be independent of the accidents of human thought and perception: the universe might be poetically imagined to resemble a river, but the accurate portrait of phenomena lay with equations that do not vary with tricks of the light, nor with the eye that gazes on the surface of the stream.

This Newtonian "answer"—the mathematical statement that can lay out an accurate account of a sequence of events—remains the target of much of contemporary science. Cosmology, for the clearest example, with its search for a unique mathematical structure whose solutions will include the big bang and the subsequent evolution of the universe, aims for a "theory of everything," as current jargon calls it. But the Nobel laureate physicist Steven Weinberg has a better term for that grail; he "dreams of a *final* theory"—final, but not complete. In his book of that title, he writes: "A final theory will be final in one sense—it will bring to an end a certain sort of science, the ancient search for those principles that cannot be explained in terms of deeper principles."

But however fundamental they may be, the scope of such first principles—what they can actually do to further the understanding of nature—is terribly limited. The proposed final theory, along with such interim steps as Einstein's general theory of relativity and Weinberg's own area of interest, the electroweak theory, does (or would) provide direct mathematical accounts of certain phenomena. Einstein's fourteen field equations of gravity yield certain exact solutions in which the radius of a spherical clump of matter drops to zero. With that, some terms in the equations go to infinity—blowing up, as the jargon has it. But these solutions do in fact describe genuine physical objects: black holes. Black holes are formed when stars collapse, forming bodies dense enough to produce gravitational fields so strong that nothing, not even light, can escape them. Einstein himself initially thought that these special solutions to his equations, called singularities, were nonsense, but the evidence has accumulated to the point of near certainty that black holes occur as a routine function of stellar life and death.

But even here the correspondence between the math and the thing itself is incredibly narrow. In its most general form, the singularity-producing solution to Einstein's field equations accounts for rotating, spherical black holes. More complicated shapes or motions might occur, but such behavior is, or would be, too complex to be modeled in the Newtonian sense of being extracted directly out of the mathematics of general relativity. And the specific problem of black holes points to the broader case: fundamental principles exist whose manipulation reveals deep insights about the physical world—but the one-to-one correspondence between the theories we devise and what we can actually observe breaks down very rapidly. General principles provide a general picture; if we want to know what is going on in detail, theories of the type Weinberg seeks will not, in themselves, do the job.

Which is another way of saying that the universe is too complicated to be rendered completely, accurately—that the loss of certainty first noticed at the turn of the last century is now, near the turn of this one, understood to be a broad feature of nature. Weinberg himself acknowledges this, writing (only slightly disingenuously) that "wonderful phenomena, from turbulence to thought, will still need explanation whatever final theory is discovered"—and moreover, that "the discovery of a final theory in physics will not necessarily even help very much in making progress in understanding these phenomena." But if final theories do not hold out much promise of help, what is the scientist interested in such (most) phenomena to do?

Baldly, such a scientist must seek something other than a final theory. Instead of trying to identify what is, she must first inquire what may be: what qualities her object of investigation has; what kinds of events could occur within it; what properties might restrict the courses such events might take. What counts as an answer shifts. The deterministic result of Newtonian science and its sequel, "final theories," tell the investigator that given certain starting conditions, this, our universe, will emerge. But where such final theories fail to reach, the answer to be sought is a qualitative one—given certain starting conditions, what kind of system will produce these kinds of behaviors; what can be found that is *like* our conception of it.

Which is where, finally, the SCID-hu mouse and the modern model come in as the most important tools of qualitative inquiry. Such models can be defined by what they are not, by the differences between them and the older conception of a scientific model. Instead of being part of nature, the essence of it, quarried out of the mass of experience and data, the new models at best resemble nature, forming images created to be mirrors of the real thing. Most significantly, a SCID-hu mouse abandons any Newtonian claim to science as a machine: modern models do not contain the complete apparatus, the connection from point to point that remains at least implicit in the concept of the clock-work universe, or a final theory of the big bang. Instead, they represent the modern form of an archaic idea, the alchemical connection between the microcosm of human artifice and the macrocosm of nature itself. The alchemists, Newton among them, had employed the oldest kind of models, Thales's type taken to the extreme, where the analogy between the image and the reality became an identity—the thing portrayed was the thing itself. Models as they are constructed now do not, of course, claim to be what they represent—but they surely do aim to imitate the real world. That imitation occurs out of sight: models take on the role of what are now termed *black boxes*, the scientist's version of the magician's top hat: a hidden chamber in which something (who knows what) happens to transform the rabbit going in into the pigeon fluttering out. Just as it does not matter to the user how a computer translates the striking of a key into the appearance of a symbol on the screen, so the mechanism that drives the behavior of the model counts for less than whether or not that behavior mimics nature's closely enough to suggest how natural processes might operate.

The idea behind this kind of black-box modeling preexists the emergence of the real thing. Hungary's Baron Wolfgang von Kempelin built or bought what he called the Maezal Chess Automaton in the 1760s, an allegedly mechanical chess-playing device. It was nicknamed "the Turk" because its moves were made by a marionette equipped with bold mustaches and a turban. A small cabinet beneath the figure held a complicated apparatus of levers, rods, and gears that apparently chose the appropriate tactics and directed the marionette's movements. The Turk played a competent game, once driving Napoleon into a rage by beating the emperor in just nineteen moves. Edgar Allan Poe, among others, recognized the deception involved, positing that the statue of the Turk hid a very small chess-playing human being within some secret compartment. (Poe, though correct, reached his conclusion for the wrong reasons. The Turk did occasionally lose games, and Poe believed that a true chess-playing machine would never err or lose. Chess computers are commonplace now, but even the most successful of them, as of this writing, still cannot defeat the very top human players in match play. Because chess offers such a wide range of possible moves, the complete theory of the game that would dictate the optimum sequence of play in every circumstance has yet to be discovered, which means that even though computers will probably surpass all human competition in the relatively near future, even the best machines will still sometimes make mistakes.)

Frauds though they were, meanwhile, the Turk and its nineteenth-century imitators demonstrated what their audiences thought ought to be possible: the creation of plausible simulations of complicated behavior. If steam-powered implements could emulate and surpass the physical prowess of those who used to weave cloth by hand, for example, so should (as Poe thought) technological inventions outwit, or at least outpace, human brains. The first genuinely successful modeled re-creation of a complex physical process, though, only emerged with the outbreak of the Second World War, elicited, like a startling number of the other scientific advances of the twentieth century, by

The Turk—a dwarf chess expert hid in the base of this eighteenth-century fraudulent automaton.

the demands of battle. By the 1940s, John Von Neumann was recognized by anyone qualified to make the judgment as one of the handful of the very best mathematicians alive. Von Neumann's scientific and political accomplishments ranged from fundamental contributions to quantum mechanics to the invention of the concept of nuclear deterrence, and thus were probably as or more significant to more people than those of any other scientist of the modern era. As biographer Norman Macrae notes, among Von Neumann's omnivorous scientific interests was the problem of analyzing complicated processes immune or resistant to penetration, either from the pure analysis of first principles or through conventional experiments. His chance to pursue the issue came in 1939, when the German blitzkrieg awoke general staffs on both sides of the Atlantic to the necessity of planning a radically unfamiliar kind of war.

Von Neumann spent most of the first years of the war working on problems of ballistics—how to get a projectile from where you are to what you wish to destroy—and on the physics of explosives—how best to destroy whatever it is you have managed to reach. But in September 1943, Von Neumann arrived at what was known officially as site Y: Los Alamos, headquarters of the Manhattan District, U.S. Engineers, home of the atomic bomb. While there he joined the team working on what had become the most intractable problem facing those trying to construct a plutonium-fueled atomic explosion.

The idea behind the atomic bomb is actually quite simple. Uranium atoms are radioactive—every now and then, an atom of the element will emit an alpha particle, the nucleus of a helium atom. In the late 1930s, a small team of German scientists discovered that it was possible to split uranium atoms artificially by bombarding them with a stream of neutrons, the recently discovered nuclear particles that possess no electric charge. Such fission releases a large amount of energy per atom. Next, Niels Bohr, among others, showed that atoms of one particular form of uranium, the isotope ^{235}U, underwent fission much more readily than the more common, slightly heavier isotope ^{238}U. Most important, ^{235}U atoms could emit two neutrons when struck by one. This meant that a chain reaction could take place—given a pure enough pile of ^{235}U, more and more neutrons would begin whizzing around, splitting more and more atoms, launching still more neutrons and releasing ever more energy with each fission event.

If the chain reaction started slowly enough and continued unchecked long enough, the result would be the release of enough heat to melt the pile—which is what would happen in the catastrophic nuclear-power–plant accident termed a meltdown. But as a number of physicists realized around the world at more or less the same time, if the chain reaction started quickly enough in a block of uranium large enough to cross the threshold of what became known as the critical mass (about eighteen pounds for the simplest bomb design), instead of a slow burn what would happen next would be an exceptionally rapid bang. Lastly, scientists in Britain, Germany, and then, somewhat tardily, in the United States, realized that all that was needed to build an atomic bomb was some method of taking two subcritical masses of fissionable material and slamming them together fast enough to set up the runaway chain reaction of an atomic explosion.

In the event, the Manhattan project actually built two different types of atomic bomb. The first used two pieces of ^{235}U, with a gunlike mechanism that fired one slug of uranium into the other. The system was easy to design (relatively speaking) and the theoretical physics behind the design predicted confidently (and accurately) that such bombs could be relied on to explode as expected. The drawback was that supplies of ^{235}U were scarce and difficult to accumulate; with enormous effort and expenditure, the Manhattan project was able to scrape together just enough ^{235}U for a single bomb by 1945. So once building the uranium-fueled weapon became a job for engineers and industrial planners, the Manhattan project physicists concentrated on the problems of building the second type of atomic bomb, one in which the fissile material was plutonium, not uranium.

Plutonium splits as enthusiastically as ^{235}U, producing enough neutrons to perpetuate chain reactions. It is a man-made element created by bombarding atoms of the heavier uranium isotope ^{238}U with neutrons. A series of reactions takes place which transforms uranium, element number 92 in the periodic table, into new substances—first neptunium, element 93, which undergoes a reaction of its own, producing plutonium itself, element 94. ^{238}U is plentiful, compared with ^{235}U, which meant that the Los Alamos scientists could count on an adequate supply. Most important, because plutonium is chemically distinct from uranium, it was (sort of) easy to separate out pure samples of the material in quantities large enough to build bombs. The only problem with the stuff was that the gun design for a plutonium weapon could not work.

The pitfall was that plutonium is sufficiently radioactive that neutrons from the first piece of a plutonium bullet would begin to interact with the plutonium target quickly, generating a lot of heat, enough for the whole assemblage to melt down before a critical mass could form and explode. So a team of the Los Alamos scientists began to work on a different technique for triggering the bomb. This second design, called the implosion method, used a sphere of high explosive to drive two halves of a globe of plutonium together quickly enough to guarantee an explosive chain reaction. Such an implosion had to be extremely precise, bringing the two pieces of plutonium together into a virtually perfect, symmetrical shape that would last the instant needed before it blew. By early 1944, it was clear that the experiments with spherical implosions were leading to a dead end. Every set of charges the Los Alamos scientists rigged produced shock-wave jets, instabilities in what should have been a smooth-wave front that were large enough to cause any lump of plutonium assembled by such a charge to fizzle out, melting into an un- or only partly exploded lump.

Enter Von Neumann. Von Neumann's great talent was his ability to recognize precisely what defined a given problem and then develop a mathematical expression of the problem that could be used to come up with a solution. He had already proposed building an explosive lens, one that would focus shock waves instead of light. By using faster-burning explosive on the outside of the triggering sphere, and a slower-burning layer inside, it ought to be possible, Von Neumann's first calculations showed, to assemble a charge that would produce a smooth, even sphere of energy that would drive the plutonium core of the bomb together.

But to actually build such a lens, the Los Alamos scientists had to know how the shock waves from the different layers of explosive would interact. To find out, Stanislaw Ulam, who would later work out a key step in the design of the hydrogen fusion bomb, took the equations that describe shock waves and began to calculate numerically what such interacting explosions would look like. No true electronic computers existed yet—the first would come on line shortly after the end of the war—so Ulam began to grind through the seemingly endless arithmetic with the aid of IBM punch-card sorters that arrived in April 1944. His calculations showed that implosion lenses could in theory be built to produce spherical shock waves that nowhere varied by more than 5 percent—good enough to produce a critical mass. Actually building such lenses proved to be an exceptionally difficult engineering task, requiring finding the right combination of different explosives and then machining the charges accurately enough to form a near-perfect lensing shape. It took the implosion team until March 5, 1945, to settle on the final design, and it was not until the Trinity atomic-bomb test that anyone at Los Alamos knew how well Ulam's prediction and everyone's work had turned out.

At that test, the scientists took bets on the explosive yield of the bomb. J. Robert Oppenheimer, director of the project, gambled (perhaps hoping to invoke the bettor's jinx) that the device would fizzle, producing no more than the equivalent explosion of 300 tons of TNT. The implosion lens builder, the chemist George Kistiakowsky, was only slightly more optimistic, picking 1,400 tons worth of bomb. One man picked zero, another went as high as 45,000 tons. I. I. Rabi, who later won the Nobel Prize for physics, arrived at the Trinity site late and took the only figure left in the pool, 18,000 tons, 18 kilotons. At 5:29 and forty-five seconds on the morning of July 16, 1945, the firing circuit in the first plutonium bomb closed. Two tons of high explosive detonated, slamming two hemispheres of plutonium together. The chain reaction began, and then observers saw and felt the light, the heat, the blast and the rising mushroom-shaped cloud. Detailed analyses later proved that Rabi enjoyed fool's luck: the bomb had yielded the explosive equivalent of 18.6 kilotons, four times more than the official estimates to which the Los Alamos laboratory had been willing to commit.

All of which meant that Ulam, working from Von Neumann's suggestion, had constructed a numerical model of a complex, nonlinear physical problem that accurately represented what would happen—that a spherical shock wave could be built to hold its shape tightly enough to produce a highly efficient nuclear explosion. He did so, in fact, with far greater precision, a far closer correspondence between the model and the event, than he or anyone else had been willing to lay money on. There are those scientists who say still that the single most notable success of numerical modeling techniques was this first one; since Ulam, the modeling of nuclear weapons has been integral to the production of bombs that became ever smaller, lighter, and more lethal. Certainly, the Trinity test and the subsequent destruction of Nagasaki with a plutonium bomb (Hiroshima was hit with the sole gun-type uranium bomb in the U.S. inventory) established the first of many precedents. From then on, scientists realized that it would be possible to devise analogs to physical processes that would reveal knowledge of the real things that neither pure theory—Von Neumann's first calculations—nor all the experiments—those explosives blowing up chunks of the Los Alamos mesa—could reveal.

The first to make use of this new method of making knowledge was Von Neumann himself. Von Neumann had observed the extraordinary tedium Ulam's group had endured, pounding through their sums on Los Alamos's hand-cranked card sorters. From 1944 forward he had acquainted himself with the design of the earliest electronic computers. After the war he set himself the task of designing and building the next generation of computers, and of using any machine he might create to tackle some fundamental problem in a radically novel way. The question he chose was how to predict the weather.

Meteorology during the Second World War had the air of magic to it, the wrong kind, as often as not. The weather forecasts for

D day, June 6, 1944, for example, were compounded out of patchy observations of existing conditions over the Atlantic, a purely historical record of typical weather for Norman springtimes, and the accumulated lore of Eisenhower's meteorologists. The aura of science that surrounded the military's forecasting efforts hardly masked the huge element of guesswork involved—as Eisenhower himself well knew. The invasion, the largest amphibious attack ever attempted, needed a considerable period of calm seas and skies to have a chance to succeed. The weather over the south of England, the Channel, and Normandy in early June 1944 proved to be unsettled, cloudy, threatening. German and Allied forecasters looked upward, consulted their data (of which, to be fair, the Germans had far less), and reached opposite conclusions. The Germans told their commander, Field Marshall Erwin Rommel, that the weather would stay chancy enough for the next several days to block the threat of invasion—so Rommel left the French front and spent June 5 with his wife in Stuttgart. Eisenhower's staff told him that he had a window of calm weather to use, and the allied commander in chief opted to gamble, making what was perhaps the most courageous command decision of the war. At Eisenhower's order, the troops embarked, and by early on June 6, launched their assault against the ill-prepared Germans.

Eisenhower, prudent man that he was, had prepared in advance a statement he intended to issue if the weather or events forced him to call off the invasion. On his side, Hitler voiced confidence that storms roaring up the English Channel would scatter the allied armada, leaving whatever portion of the armies that had made it ashore prey to a defeat worse than Dunkirk. The storms that did, in fact, batter the invading forces in the weeks after D day demonstrated how close Hitler came to being right.

Von Neumann thought there had to be a better way, some method of calculating what the weather would do next, instead of relying on sightings and anecdotes. The idea of constructing a numerical model of the atmosphere, within which one could simulate with arithmetic the behavior of the atmosphere, had actually occurred to the British meteorologist L. F. Richardson shortly after the First World War. His attempt failed—he chose the wrong mathematical expressions, and in any event, he lacked any machine that could do his sums fast enough to keep up with the pace of change out in the real world of wind and rain and sun.

Richardson eventually imagined creating an amphitheater that could hold 64,000 human beings, each grinding away on a hand-cranked calculator, kept in sync by a kind of orchestra conductor waving a baton in the middle of the arena, coordinating the incredible volume of arithmetical operations needed to produce a timely weather prediction.

Von Neumann set about overcoming each of the obstacles Richardson had encountered. He got the theoretical meteorologist Jules Charney to construct a series of increasingly complex mathematical models of the atmosphere, beginning with one that assumed the atmosphere was a uniform two-dimensional sheet, in which events moved north and south, east and west, but not up and down. More complicated versions allowed winds to blow in the vertical direction, exchanging energy, encountering differences in pressure, and so on. Meanwhile, Von Neumann and Charney worked out the formal language needed to convert the pure mathematics of atmospheric physics into numerical expressions that a machine could add or subtract, divide or multiply. Finally, Von Neumann directed the effort to produce a computer that would have the calculating power of 100,000 people, clearly beating Richardson's imagined arithmetical orchestra.

The first Von Neumann–Charney predictions left a bit to be desired. Using the older, slower, original ENIAC computer, they did produce one good forecast of the next day's weather—but it took them thirty-six hours of computer time to do it. Another trial run failed completely, producing the sort of result that would have cost General Eisenhower what hair he had left had he relied upon its forecasts. But by 1952, the team could generate weather predictions with just ten minutes of computation, and by 1955, computer-model forecasts were consistently as or more accurate than human predictions for as far ahead as two days in the future.

Von Neumann believed that the next logical step would be to extend such forecasts, developing within the computer an increasingly sophisticated and accurate mathematical representation of the atmosphere that could predict the weather indefinitely far forward in time. But Edward Lorenz's famous "butterfly effect" killed that hope. Lorenz's model-experiments on the stability of the weather proved, in effect, that the atmosphere has a short memory. After some time, calculated to be about thirty days, the atmosphere as it is holds no clues to what the atmosphere was like a month ago. Beyond thirty days, that is, detailed weather prediction becomes impossible.

Nevertheless, Von Neumann's weather model was the grandfather of the fundamental tools now used in every significant weather forecast in the world. Like Von Neumann's original effort, contemporary models attempt to re-create within a computer as realistic as possible a mathematical description of the atmosphere, in order to generate what appears to be a deterministic forecast. Given a set of starting data, that is, the models try to say that within three days (for example), the storm system brewing up at the intersection of polar and Gulf of Mexico air masses will dump between eight and twelve inches of snow from Bridgeport to Boston. But because the models are actually only representations of the atmosphere, such predictions are not deterministic—no one can be absolutely sure that the eight inches of electronic snow that pile up inside the computer will actually fall on Beacon Hill. Rather, the predictions are probabilistic—the weatherperson on television will announce an 80 percent chance, perhaps, a figure that is a kind of measure of how much of nature itself is thought to have been captured by the portrait of nature written into the model.

The collateral descendants of Von Neumann's original model extend the reach of pure forecasting models—simulations exist to re-create ice ages, or to examine what might happen if the map of the continents were redrawn to alter the circulation of currents within the oceans. Physicists have built models to probe the behavior of subatomic particles, governed by the complex relationships of quantum chromodynamics, a subset of the quantum theory that examines what happens inside the nucleus of the atom. Models of fluid dynamics simulate the flow of air over wings and fuselages; planes plummet many times within the computer well before any new form flies at Edwards Air Force Base. (Edwards, incidentally, was named for a pilot who died in the crash of an early flying-wing design. That plane was uncontrollably unstable, which, in the absence of a valid model of airfoil behavior, became obvious only when the full-scale prototype flew and fell. The flying-wing idea was finally sucessfully employed for the B-2 stealth bomber, a plane designed with the aid of computer simulations.)

The unifying constant in all the models developed for either basic research or technological development is that they combine rigorous physics, chemistry, biology, or what have you, with what can be understood as a form of fiction, a kind of storytelling. The rigor imposed by a model's underpinning of scientific laws shapes its tales, as does the ultimate corrective of actual observations. All model runs can be compared with what happens: the hurricane that does or does not make landfall on the south coast of England; the oil found or not whose existence was predicted by a seismic model; the plane that flies uneventfully (one hopes) from San Francisco to Sydney. But as the failure (still! thirty years and more after Von Neumann first confidently tackled the problem) of weather predictions suggests, simulations of complex systems are not the systems themselves; the models are not the world. Strictly speaking, the only information a model provides is data about the model itself.

The ultimate example of the solipsism inherent in a science that depends on models comes from a field on the far edge— some would say the fringe—of contemporary research. In his book *Artificial Life*, Steven Levy has chronicled the efforts of a loosely interlocking group of scientists who are attempting to devise simulations of biological processes that display, in their electronic environments, the characteristics of true, living organisms. Levy's account begins with the story one group of creatures, differentiated by color, that seek to survive. They receive signals from their world—they know that if they encounter a blue creature it is looking for a mate, while red ones are ready for battle. They know themselves a little: they can sense their own hunger, their weariness. They can learn. They breed. The intermingling of the information needed to make new creatures, genes coded in computer language rather than DNA, permits evolution. New forms of the creatures emerge, species differentiate: edge dwellers appear in a unique niche in the ecosystem, while sometimes a species nicknamed the "cannibal cult" flourishes by living and mating with (and dining on) its own members.

All this happens on the screen of a computer workstation, running a program called PolyWorld, written by Larry Yeager of Apple Computer. With every run, Yeager sets PolyWorld in motion, but he has not programmed the end result; he does not know which of his creatures will finally survive. He lays out a set of simple rules that governs interactions among the creatures. Then he sits back and observes the variety of complicated behaviors that explode on his screen. To Yeager his entities have a status that extends beyond that of simulations of living processes —they are, in fact, a new form of being: "artificial life"—as Levy titled his book.

Life? In a computer? Yeager's is, certainly, the most extravagant claim that can be made for a model. He asserts that the lifelike behavior of his creatures or the other products of artificial-life (or a-life) research implies that the processes that underlie the model's behavior are among those that govern living things beyond Yeager's workstation. According to such bravado, the model has some direct correspondence to organic reality. But by calling his creatures "artificial," Yeager comes close to retracting such a thought: the quality he claims for his creatures—that they live—he conditions, he limits: they do so as artifice. His model, he thus admits, like all models, reveals truths that hold absolutely true only about itself. Its entities exist (really), but they do so within the structure of his programmed, artificial world.

That need not matter much, if the goal is simply to study the PolyWorld creatures on their own terms. But the makers of most models seek to establish a connection between their results and nature. The global AIDS epidemic provides the clearest contemporary example of a problem too complex to be reduced to a simple task of precision measurement and calculation. The SCID-hu mouse is one of the most promising tools being developed to confront the disease. That mouse is a model—a little unusual perhaps, in its use of a living organism rather than a computer. But from its first design to its application to the problems of AIDS, the history of SCID-hu development illustrates each of the hurdles the new model-based science must overcome to produce knowledge that can render intelligible some portion of the irreducible complexity of the world.

The path that led to the SCID-hu mouse began with a single, critical, and in some sense arbitrary decision. "A model should be made to look 'like' something else," wrote Mike McCune, SCID-hu's inventor. "In this process there is choice: we must decide what it is we wish to mimic." McCune made that choice in 1987, while performing postdoctoral research at Stanford. His second job, as a physician at an AIDS clinic, forced him to recognize how little was actually understood about how the disease actually progressed. What he needed, McCune realized, was a copy of a working human immune system—a simulation of immune function through which he could observe AIDS in action, a model whose form he could control and whose behavior he could manipulate at will.

The SCID-hu mouse he built uses the same basic logical structure all simulations employ, whether they employ a computer or as exotic a technology as a hand-modified mouse. All simulations work by abstracting from whatever it is they seek to describe a simplified set of functional subsystems that can reproduce at least some of the original phenomenon's behavior. Von Neumann with Charney used a set of equations that would account for most of the atmosphere's motion. McCune worked with a few human tissues, chosen out of the much larger collection of organs that take part in immune function. All models require some engine that runs the processes being simulated. Von Neumann built a computer and had to write his equations in a form his machine could process. McCune used a mouse, within which his human tissues could grow and function, and he too had to select his particular human components carefully to ensure they took a form that could safely inhabit a different species. All models need to be validated. Von Neumann's successors, McCune included, have had to test their models, first against their own design expectations and then against the much harsher standard of the real world. Finally, all model builders have to use their models to discover something new—otherwise the entire effort is a waste of time, an exercise.

In other words, McCune faced four tasks as he set out to create a model that would help him understand and treat AIDS. First he had to work out a detailed design—which meant that he had to build the simplest possible immune system that could still work, still provide a healthy immune response to invaders. Second, he had to prove that his design did work—that the immune system he built into his mice actually protected them. Third, he had to go beyond the model itself and show that the immune system contained within the mouse actually behaved more or less like a true human one, in situations where the outcomes were well enough known to provide a clear standard against which to judge the performance of the model. Fourth, he had to make a prediction based on some new event he observed in his model that he could test against the real world. To be worth the effort, the mouse had to produce new ideas, ones that matter for the treatment of AIDS, concepts that could not be found either by looking at human patients or through test-tube experiments.

The general problem in dealing with the first step, design, lies with finding the right way, the right language, to translate a complicated reality into useful, functional abstractions, simplified versions of the real thing. For mathematical models, the question is how to express the formulas in a way that can be solved—and that can be a real problem. Some equations cannot be written so that they can be easily resolved to numerical calculations. For the SCID-hu, the task was to figure out how to match human components to a mouse body in a way that both would survive. Irving Weissman, McCune's boss at Stanford, thought the job impossible, and he reminded the younger scientist that others had already tried—and failed—to transplant human immune tissues into other species. There are two major obstacles to such transplants, as both Weissman and McCune knew. First, the host animal's immune system might do what immune defenses are supposed to do: recognize the human tissue as foreign and reject it. If that didn't happen, the human immune transplant in its turn could function too well, registering the animal in which it found itself as nonself. That would trigger an assault by the human cells against the surrounding animal organs—a painful, wasting, deadly syndrome called graft-versus-host disease.

McCune, though, had a double insight. His choice of a SCID mouse offered him a way around the first difficulty. Lacking immune function of their own, SCID animals could not reject any tissue McCune chose to implant. To overcome the other barrier, McCune proposed the truly radical idea in his model design. Instead of mature, functional immune tissues, he decided to try implanting fetal human grafts. Fetal immune cells are not yet immunocompetent—they do not make the critical distinction between self and other that creates the immune response to infection (which is what keeps a fetus from rejecting its mother; the hormonal changes of pregnancy modify the mother's immune response to prevent rejection in the other direction). As fetal tissues mature, they encounter all the proteins that make up the organism of which they are a part—and all those molecules are catalogued as self—while any new, unfamiliar proteins provide the signal that anything attached to the protein (like a virus, a bacterium, or whatever) is alien, to be attacked. In theory, and, McCune hoped, in practice, fetal tissue would mature as happily inside a mouse as in a person and would list both the host mouse and its own human proteins as self.

With this broad strategy, McCune selected the specific human tissues to implant. He chose just two different types, hoping that such a radically simplified combination would still generate immune function. The first, liver, would produce human blood cells, including the cells that generate an immune response. The second, thymus, is the organ in which certain cells learn to distinguish between self and other. (Adults make new blood cells, including all the cells of the immune system, in their bone marrow, but before fetal bone marrow matures to the point where it can take over the task, blood-cell formation, called hematopoesis, takes place inside the liver—which dictated McCune's options.)

The process of actually building the mice was relatively simple. For the first batch, McCune, now joined by four others, harvested liver and thymus from fetuses of about nine weeks gestational age. The team implanted thymus into eleven SCID mice, keeping ten more as a control group. A week later, they injected liver cells into the test group of mice. Then they waited. Step two, proving that the model worked, at least as far as the mice were concerned, came in two stages. First, blood taken from the eleven implanted mice, now officially designated SCID-hu, started to reveal populations of distinct, differentiated, clearly human immune system cells. That meant that the human liver had actually matured within the mice to the point that it could produce new blood cells, and that the thymus graft was in fact teaching those cells their immune function. From the very first, then, the model "looked like" what it aimed to copy—at least a bit. The second proof was both less sophisticated and absolutely unequivocal. Of the whole mingled colony, the eleven modified mice lived, while the ten controls succumbed to pnemocystis pneumonia, an opportunistic infection that homes in on immunocompromised individuals. The survival of the eleven SCID-hu mice suggested that their human components did more than just appear to function. Instead, they actually made a difference, providing the modified mice with protection against disease that the others lacked.

All of which meant that McCune's basic idea worked—but no more. He knew that his mice had a functional immune response of human origin, but he had no proof that those systems behaved in the mice in the same way that healthy, complete immune complexes work in human beings. Step three—proving that any given simulation is not just a copy but a convincing mirror of nature—took the form (as it usually does) of asking a question of the mouse to which the answer for people was already known. Again, this test took place in two parts. First, the now growing SCID-hu research group tried to infect their mice with human immunodeficiency virus (HIV)—the agent believed to cause AIDS. In their initial trials, the protocol was as crude as it could be: the team simply injected large doses of purified virus right into the implanted thymus. This brute-force effort had the desired result: at the right doses, every mouse they injected tested positive for HIV infection.

With that, McCune's group attempted the second part of the proof. Clinical trials on human patients had already shown that the antiviral drug AZT could alter the replication, or spread, of HIV in the body (though AZT's impact on the disease course of AIDS itself is still less well understood.) Further tests had come up with the optimum drug doses to create the greatest chance of benefit. To compare their animal model to the human experience, McCune's team gave their HIV-infected mice a range of different courses of AZT. In a major confirmation of the accuracy of the simulation, they found that the optimum doses, the ones that most slowed the spread of the virus, were the same for mice as for men (with the appropriate correction for body weights).

By 1990, McCune and his team were in the same position Von Neumann had reached in 1952, midway through the development of his first atmospheric model, with which he successfully "predicted" the weather for a day two years in the past. Such a success did not count until the Von Neumann team could actually make predictions in the other direction, into the future. Similarly, McCune's group knew that valid and useful as their model seemed, they still had to demonstrate that SCID-hu would actually lead them to unsuspected insights about HIV and AIDS, about immune failure and health.

More drug trials provided the first hints that the mouse could anticipate discoveries made in human populations, though an ongoing lack of faith in the meaning of model results meant that they were largely ignored. Dose-level studies for DDI, another antiviral compound, were still under way in human patients in 1993, with the results converging on a number predicted inside SCID-hu mice in 1990. The story of another compound had a sadder outcome. HIV binds to a particular site called CD 4—a protein structure that protrudes above the surface of immune system cells. A number of different groups and drug companies reasoned that synthetic CD 4 proteins could be made, using recombinant DNA techniques, which could then act as decoys, binding with HIV, preventing the spread of the virus throughout an infected patient's body. In the test tube, the idea worked; HIV bound readily to the manufactured molecules.

Given that boost, human clinical trials began, while at the same time, McCune's group started to test the drug on the SCID-hu system. Preliminary results came from the mice first:

nothing was happening; CD 4 did not appear to alter the course of the HIV infection. McCune's team raised the dose—and still nothing changed. McCune's first thought, given the success of the test-tube experiments, was that the model simply was not good enough, that it was too sketchy, too incomplete an imitation of human immune function to handle the complicated biochemistry involved in the CD 4 mechanism. As it turned out, though, the model results were perfectly in tune with the clinical outcome. The human trials begun with great hope and considerable publicity ended quietly after about a year, without success. All clinical trials are a gamble, but the hard truth of this particular failure was that the SCID-hu experiments offered a clue about the probable outcome of the test—and the patients enrolled in the experiment lost about a year to what turned out to be a worthless drug. Belief in the mouse could have prevented that waste—but as McCune's own initial reaction demonstrated, the SCID-hu model was still too young and too unproven to be persuasive.

The breakthrough came with an investigation of the fundamental process that lies behind the creation of healthy immune function. Hematopoesis, the production of new blood cells, begins with a single type of cell. That ancestor, called a stem cell, divides into daughter cells that differ from each other, granddaughters that differ further, and so on, until the original stem has generated all the branches that make up a complete blood system, including the cells that create immune function. McCune's old boss Irving Weissman, among others, had gotten as far as proving that stem cells exist as only a very small percentage of bone marrow, and that such cells display certain biochemical markers. Some blood cells will grow and differentiate in tissue culture, but no one had successfully raised the complete list of blood-cell types in the test tube, so no method existed to test the claims of any candidate stem cell. Enter the SCID-hu model, specially modified for the task.

By 1989, McCune's group's increased familiarity with the behavior of fetal human tissue within mice had led to a number of changes in the model. In the original SCID-hu mouse, the "hu" portion of the model seemed to wear out over time. The tissue grafts would produce an initial wave of human immune cells that wouldn't be replaced by new cells, generating an immune response that would fade after a few months. A series of experiments with different immune tissues—lymph node, spleen, skin—led finally to a system which used actual liver tissue, rather than just cells in solution, implanted in a single lump with thymus, producing a strange, hybrid organ McCune's group dubbed (prosaically) "thy-liv." The new invention solved the original problem: thy-liv mice generated the original, full wave of human blood lineages—and kept on doing so. Critically, the new SCID-hu mice produced a full population of T-cells, whose disappearance marks the progression of HIV disease toward full-blown AIDS.

With this success, the new version of SCID-hu laid bare the central tension of model science, the line that stretches between what models reveal and what they hide. The thy-liv organ exists nowhere in nature. It cannot be a copy of any "real" natural phenomenon; it does not reproduce the parts of the original system, and whatever it does, it cannot be doing it in the same way that human organs inside human bodies perform their functions. It is truly a black box. McCune and his group did not know when they built it how it might work; they still do not. They only have understood, since it began to operate, that work it does. Such knowledge is sufficient, such ignorance survivable. In the context of the model, in the hope of understanding AIDS, the mechanism does not matter: results are all. Thy-liv contains mysteries. It also makes blood.

To locate the stem cells amid that rich harvest of blood, McCune and his colleagues began with cells harvested from human bone marrow. They subdivided the sample, using a machine that can distinguish between cell surface markers. Cells that clearly belonged to already distinct daughter lineages were eliminated at this stage. Tissue culture tests eliminated some more types. Then the team took fractions of the remaining blood cells and injected them into SCID-hu mice. Those mice that produced all the different types of human blood cells had received some stem cells. To figure out which of the remaining cell types in the sample was the one true grail, the team simply repeated the procedure again and again, giving SCID-hu mice smaller and smaller fractions of blood. By 1992, when the team had managed to restrict fractions to what were clearly single types of cells, they knew that those mice that generated the complete catalogue of human cell types must contain the genuine human hematopoetic stem cell.

As expected, such cells are as rare as can be: they account for no more than .01 to .05 percent of the total bone marrow. With them in hand, McCune could (at last) attempt to connect his model studies to real results in the clinic. He looked first to cancer rather than HIV. Certain forms of the disease, notably breast cancers, spread readily to bone marrow. The treatment for both breast and bone-marrow tumors includes radiation and chemotherapy—both of which are dose-limited: too much and they kill the marrow, and ultimately the patient herself. Bone marrow transplants can replace what such procedures destroy, either with marrow harvested from the patient herself before treatment or with tissues from some other donor. But when a patient receives her own bone marrow back she can get her cancer returned along with the package, while bone-marrow transplants from other donors always pose the risk of graft-versus-host disease—and in any event, both operations are exceptionally painful.

The ability to isolate stem cells offers (in theory) a way out. McCune and the company he founded to develop SCID-hu technology began in 1993 to design a clinical procedure that would begin with the harvest of bone marrow from advanced breast-cancer patients. The next step would be to isolate the stem cell population in the marrow, testing it using SCID-hu mice, and checking to ensure that none of the cells have become cancerous. Then, at the point when the cancer sufferer would usually require a marrow transplant, they intend instead to try the much simpler and less painful procedure of injecting stem cells back into the irradiated bone. If everything works, those stem cells would regenerate a complete, healthy blood supply, free of both cancer and the risk of rejection.

Such a trial would be the final validation of the SCID-hu system. Making a human blood supply inside a mouse may come close to proving that the model faithfully replicates the original. But as far as McCune is concerned, the case will not be settled until cells taken from some person, checked out in SCID-hu animals, and replaced in a human being actually act as stem cells must. As of this writing, that last milestone has not been passed. But in the hope that it will be, McCune has already identified a way his mouse could lead him to his original desire: some kind of victory in the treatment of AIDS. The outline of the idea is simple enough. Many groups have found genes that confer resistance to HIV infection. Techniques exist to transplant such genes into cells that lack them. Put those two results together with SCID-hu, McCune argues, and the way lies open (apparently, maybe, with the greatest of good luck) for an end run that could neutralize HIV disease. In practice (or rather, still, in theory): begin by harvesting and isolating stem cells from someone infected with HIV. Check to make sure that (as currently believed) the virus does not attack those particular cells directly. Implant one or more genes that prevent HIV infection into those cells, and then grow, either in test-tube cell "factories" or in SCID-hu mice, as many copies as possible of the genetically altered originals. When enough stem cells have been made to supply a full-sized human need for new blood, transplant the HIV-immune colony back into the infected individual. All the daughters of the new cells would continue to carry their new genes; a new complement of HIV-resistant immune cells would emerge (presumably, hopefully), all of which would persist, even in the presence of the virus. HIV reservoirs could remain inside someone, probably forever. But the progression from infection to disease would be broken: a new, healthy immune system would exist, operating alongside the ruins of the old, HIV-ravaged one—or so McCune and everyone involved in the AIDS epidemic deeply desire.

As of 1994, such ideas are merely dreams, imaginary tales. Others like them have ended in bitter disappointment throughout the years in which AIDS has continued to spread. Clinical trials may begin as early as 1994, but any results will not come in for years. Even if McCune's idea works it will not represent a cure for AIDS, for it will not affect other aspects of the pathology of HIV disease, including the dementia caused by infections in the brain. And there is a growing body of work that suggests that other factors besides HIV can contribute to the destruction of the immune system, which, if true, will increase the complexity of any treatment for AIDS. Yet should the large group now pursuing McCune's original idea discover a way to reconstitute healthy immune systems, reducing or removing the threat of the opportunistic diseases that destroy the lives of most people with AIDS, this line of research would represent by far the most significant advance yet in the battle against the epidemic. AIDS would retain some of its long-term consequences, perhaps, but if (that dangerous word) McCune's idea pans out, the disease could become (as today's slogans so bravely proclaim) one with which people would live.

This is the end to which the SCID-hu pointed from the first, one example of a kind of science new to our own time. Its accomplishments so far are entirely typical of the form. It makes no assertion of mechanism—the pieces that make up the model do not correspond, one for one, with those of its natural counterparts. It presses instead a claim of resemblance, of mimesis. SCID-hu, all models, are looking glasses. The world does not reside in there, in the model. But it can be glimpsed, caught in its parts, constructed—or rather reconstructed—by the mind that examines the mirror. SCID-hu told its users that there was a stem cell; this is what that sought-after object looks like; here is where it can be found. The proof lay not with a dissection of the rules that govern such cells' function. Instead, it was marked by what happened, by the fact that a mouse, suitably equipped, could use that one thing and no other to build what looks like all you need to make a human being's blood.

Looking at a photograph differs from looking at the subject of the shot. In today's science we look at models because we have to, to meet a pressing need for knowledge, to construct knowledge that we cannot seize in any other way. The older, Newtonian-derived idea, with its metaphors of the machine and its enormously successful methods, did and will yet achieve remarkable results. Looking directly at the disease, at HIV, at the dynamics of the information, research into AIDS has come up with more information in less time than ever before achieved for any human ailment. HIV's genes are known. The proteins that cloak it have been characterized. Other viruses associated with the progression of the syndrome have been found. The sequence of events that leads to the collapse of immunity has been traced in part. The life cycle of HIV has begun to unravel. These are facts, observations, taken directly from nature; these are things we know.

But the AIDS epidemic, for all that it has yielded to the traditional scientific method, still continues; the disease spreads, people die. The reductionist approach of looking directly at the chain of cause and effect in AIDS has led to a picture of the process, but not to the magic bullet. The decline of a human immune system seems to be less a mechanical failure than an ecological collapse, a multiple catastrophe too tangled to resolve (so far) by fixing this part or that. The metaphor of the machine, that is, has taken the investigation of AIDS into a labyrinth, in which every step forward reveals more of the maze—but does not necessarily take us any nearer our goal.

The use of a model creates a shift in the image, in what the modeler sees when he considers AIDS—and it is this change in perspective that is the model's most important contribution. SCID-hu emphasizes the system, not the specific elements that either make it work or break it down. The solutions that such images evoke do not limit themselves to some disruption of the virus or a new drug for an opportunistic infection. McCune's plan to use genetically altered stem cells to build better immune systems may or may not work. But it represents the type of response required by the complexity of the biological process under scrutiny. The model with its created immune system carries with it—built in—the reminder that the goal is to restore immune function, to rebuild the whole, rather than merely to halt disease, to slow disintegration.

Here again, a model gives a hint not of how to control nature but of how to adapt to it. More generally: models allow us to see what could be real, and then permit us to form our judgments accordingly. Such knowledge requires of us the decision McCune spoke of, the choice of what to imitate. We can pick whatever pleases us, even the surreal view of a mythical chimera made flesh, both mouse and man. But we cannot choose all we want. On the one hand this is a simpleminded truth: when you look in one direction, other vistas pass out of your field of view. A telescope staring at a star orbiting a black hole reveals nothing of the formation of spindles that takes place when a cell divides, and no one expects it to. More deeply, though, the use of models demands a second layer of choice. First you select something to imitate; then you must determine, create, imagine the meaning of what your imitation reveals. Models emerge because of the inherent uncertainty built into the human understanding of the complex reality of nature. They offer a way to hem that uncertainty in, to define it, to fix our attention on the aspects of nature that are intelligible. They do this by pointing to behavior, to the potential range of action of any system. But in doing so they retain (as they must, for they are, after all, imitations, copies) their portion of uncertainty. At the extreme, misconstructed or misconstrued models generate not meaning but farce—and there are plenty of them around. Consider the (minor) daily errors in weather prediction, or the more serious perennial prediction of the collapse of the world's oil supply, derived from a number of ill-conceived, multifactorial models, all of which—so far—have proved false. The worst of model science confirms what is valid for the best as well: we cannot be sure of what we think our models have revealed to us because they require subjective, fallible judgments to make them go. No law tells us which questions to ask, nor what value, what meaning, to attach to the answers we find.

How to decide what to do when all choices are open and how to make sense of what you have done is the ultimate challenge for any inquiry. How that challenge plays out, and what can be done about it, is the subject of the next chapter. For now, there is a story told from the beginnings of the history of scientific thinking that expresses, better than anyone has since, what this new form of inquiry attempts, and what it has abandoned. At the opening of book seven of the *Republic*, Plato tells the famous parable of the cave. First, inhabitants of an underground den, chained since birth, can only stare forward, at the wall of the cave. Behind them, out of sight, a fire burns, and as the prisoners sit or walk along their narrow promenade, they can see only the shadows cast in the light of that fire. As Plato's narrator, Socrates, concludes here, to the prisoners "the truth would be literally nothing but the shadows of the images."

Next, Socrates said, imagine what would happen if the prisoners were released. As they turn, the glare of the fire pains them, and they have no words, no categories of thought with which to identify the objects now illuminated and visible, whose shadows alone they had known before. In the pain and confusion of new sights and thoughts, Socrates suggested that a prisoner would find "that the shadows which he formerly saw are truer than the objects which are now shown to him." Finally, once eyes and minds have become accustomed to the pale illumination of the fire in the cave, consider, Socrates proposes, what would happen when some hardy, brave soul ventures out of the den altogether and looks at the world in the light of the sun. He would then again have to overcome the pain and confusion of true illumination. He could begin by glancing into ponds to observe reflections of what is. Then he could examine the objects themselves, the stars and the moon. At the last he could stare directly at the sun, the true, eternal essence of things.

As the parable comes to a close, Plato allows Socrates to predict his own bad end. A trial and the sentence of death would be, Socrates said, the inevitable fate of a truth seeker who finally penetrates all the way to the ultimate reality, the Good—and then has the gall to tell his fellow citizens about it. But the twin beliefs enshrined in the story survived the deaths of its principal characters, forming the pillars of scientific faith. The first doctrine of the faithful was that the truth exists, as Plato asserted, a single, absolute, eternal, unalterable reality. The second holds that this ultimate truth can be sought and found. Its elements can be deduced from the shadows, the accidental objects we may see, the reflections that reveal one face or another of the ultimate reality. Human understanding can apprehend all this ordinary experience, the material substance of the world. And if the thinker can conceive of a final truth behind all the apparent reality he encounters, then all these elements derived from that one source will lead the thinker upward, into the full light of the sun.

In all of its forms, almost to the present day, this is what science was about: the search for an ultimate truth that proceeds through deeper and deeper investigations of the perceivable world. The definition of an ultimate truth changed, from Plato's "Good" to Weinberg's "Final Theory," but the underlying vision of a light shining beyond the mouth of the cave abided. The Newtonian revolution ended and ours began when that vision failed. The discovery of uncertainty in all its forms casts us back into the cave. The invention of methods to confront uncertainty represents an adaptation to life in the dark: instead of trying to see past our den, we now try to find our way within it. The use of such expedients, models that act like the mirrors Socrates rejected as inadequate, transforms what science is about: not a single truth, but the discovery of many truths, partial accounts, versions accurate enough to propel us on to the next telling of the tale.

In his brief book, *Invisible Cities*, the Italian writer Italo Calvino tells a fable that speaks across time to Plato. Calvino places a young Venetian explorer named Marco Polo at the feet of the Mongol emperor of China, Kubilai Khan. The Great Khan commands Polo to tell him of all the cities he has encountered within his realm, cities that Kubilai himself will never see. Marco speaks, reporting on place after place, but as dawn breaks he stops, and admits that he has completed the stories of every city he knows. Kubilai corrects him: "There is still one of which you never speak . . . Venice." Marco responds, smiling: "What else do you believe I have been talking to you about? . . . Every time I describe a city I am saying something about Venice."

The answer vexes the emperor: "When I ask you about other cities, I want to hear about them. And about Venice, when I ask you about Venice." It is a truth he cannot have, according to

Marco, or rather, it is one Marco cannot find, or make. "To distinguish the other cities' qualities I must speak of a first city that remains implicit. For me it is Venice." In Calvino's telling, Marco does not return to his home city, nor does Kubilai ever get the account of Venice that he seeks. The emperor—we—can only examine the reflections of all the images that could be created, each containing something of an echo of the whole, the single city from which all others derive. Within the emperor's garden, within the cave of human experience, Marco—we—may only recall what we have seen, and speak of it, then journey again, see more, speak of it again, and journey once more, around and around and around, until we have seen enough.

Chapter 8

WE NEED NEW

INSTRUMENTS

VERY BADLY

The building sticks out amidst its drabber neighbors. Standing on the edge of the Massachusetts Institute of Technology campus in Cambridge, Massachusetts, the blocky, monolithic, I. M. Pei–designed cube seems a comic, almost rude gesture directed at the ever-so-proper brick-faced laboratories around it. Whatever boring and conventional work might be going on in such places, the cube mutely suggests, the action that matters happens here. Certainly that is the boast that some of those who inhabit the building have made. The cube houses MIT's Media Lab, where, as the local slogan puts it, the laboratory's researchers are engaged in nothing less than "inventing the future."

There are corners of the lab, though, where the future seems to have gotten itself thoroughly tangled in bits and pieces of a tenacious past. Up on the fourth floor a visitor may find state-of-the-art holograms on display in the printer room, or might find himself the subject of an experiment on how one might actually interact with interactive television. But down in the basement, in a long, narrow, windowless corridor of a room, technology more than a century old collides with machines—and ideas—on the edge of the new.

Clutter dominates the space, which belongs to a young man named Michael Hawley. Hawley has spent the last several years teetering on the edge of earning his doctorate, finally leaping the last hurdle in 1993. Over time, the layers of Hawley's life have steadily accreted within what is nominally his office to the point where the distinction between home and work has almost disappeared. His canoe hangs from the ceiling. His ski gear leans against one wall, while several portraits of Julia Roberts smile down from another. A couch of no particular design or color serves by day as Tasha-the-German-shepherd's bed, and looks as if it occasionally does double duty for Hawley as well. There is a gem, though, almost lost within the thicket of stuff. Stretching almost out of sight into the ill-lit back corner of the room stands Hawley's Bösendorfer Imperial concert grand piano, the largest (and some say the best) piano commercially available. It extends nine feet, six inches; possesses nine extra bass notes, for a total of ninety-seven keys; and weighs almost three quarters of a ton. On a concert stage, the Imperial appears as more than just a musical instrument: it becomes something of a work of sculpture, a monument to the idea that where it is, music happens. Hawley's version, though, looks a little different. The clean black lines of the instrument have been broken up by the addition of a large, silver-colored metal box that stretches the width of the underside of the piano just behind the pedals. The box holds a Kimball digital playback and recording mechanism—an automated piano control system. The Kimball device uses lights to measure the timing and speed with which the Bösendorfer's hammers strike its strings. The Kimball then transforms the flicker of the optical sensors into the digital data that computers can read. Finally, it plays back its digital musical information by activating a stack of electromechanical switches that can drive the instrument hammers and keys.

Put another way: Hawley's concert grand piano is, in his words, "the most magnificent player piano in the world." Hawley even has a paper-roll mechanism ready to hand. Made of Lego blocks, and driven by a tiny electric motor, the device unravels a paper scroll in front of a video camera. A computer, using custom software designed by Hawley, can convert the punched holes of a player-piano roll into the digital signals to be transmitted back to the solenoid switches that command the Bösendorfer's keyboard. Scott Joplin's rags, Rachmaninoff's pyrotechnics—if they were punched onto paper, Hawley can (with some effort) capture them and set them free again, booming out in that magnificent Imperial sound. Hawley's system, the piano, its associated computers, the paper-roll reader, cobbled together out of children's toys and modern electronics—all this together becomes the incarnation of a century-long development of an idea: a change in the way one can conceive of music, in the way one may imagine what music is made of or from.

Hawley's instrument, though, stops one step short of completing that process. Whatever happens inside its electronics, it will always sound like a piano. Other instruments are not so limited. Five floors above the Bösendorfer's basement lair, a group of researchers led by the composer Tod Machover has constructed one version of the device that completes the logic of the development that leads to and through Hawley's machine. Called a hypercello, it is a customized extension of the ultimate musical device: the digital sampling synthesizer. Such synthesizers are universal machines, capable, in theory, of representing any sounds as strings of bits, ones and zeros. From the begining, with the organ through to the player piano and finally to the synthesizer, the evolution of the technology of musical instruments has both reflected and driven the transformation of music. Where it once was only an expression in sound—something heard—in our century music has also become information, data—something to be processed.

There is a problem with such a concept, one with which contemporary science is familiar. Twentieth-century science confronted the problems of complexity with the systematic use of simulations, models that imitate rather than analyze the elements that make up the world around us—a strategy that has succeeded in a growing number of realms. But the use of models contains a pitfall. Given that all scientific simulations are incomplete, that all require some element of subjective, personal choice in their construction, and that all produce results that need to be extrapolated, interpreted to make the leap from the model world to the real one—why then do models work at all? Even worse, because a model can be built to test any assumption, because there are an infinite number of imaginable simulations that one might run, how do we know when we have learned something meaningful about the world? What must occur, what properties must exist in the universe and in ourselves to allow us to distinguish between fantasy and some persuasive account of fact?

In music, remember, Stravinsky comforted himself by noting that he was hemmed in by the twelve-tone scale and the particular range of sound his orchestra of instruments offered him. Today, though, neither the musician nor the scientist can take refuge any longer in such a clearly defined sense of what the universe contains. The invention of the synthesizer makes the problem for musicians explicit, and identical to that of the scientists: When all sounds are available, how does one make sense,

make music, out of this ocean of possibility? The story of the
synthesizer—where it came from, how it works, how it has been
used—provides an outline of one set of answers to such questions. By extension, by analogy, the sound of music evokes (as it
has before) an instance of how science works, how we manage to
impose a sense of order on the limitless range of our experience
of the world.

The synthesizer itself was born of the electronics and semi-
conductor revolutions of the last two decades, but the roots of the
instrument lie (of course) much deeper in the history of music
making. Music was first, and remains for most, sound made and
heard. Gregorian chant, memorized note for note and word for
word and sung at all the Offices of the day, existed aloud, as the
song actually sounded in church. Music—any music, from chant
to pop—existed and persists in and through a performance, the
production of ordered tones, listened to by an audience. That is
almost a commonplace: music happens in the real world, in real
time, as a listener experiences the sensation of sound, as she
actually hears what she or a loudspeaker or some other performer
produces. But almost from the moment Western culture began to
reemerge from the collapse of Rome, the invention of techniques
of writing music suggested a different conception of what constitutes music. Musical notation records in symbols the fundamental variables of musical expression: the choice of instruments
to produce the desired characteristics of sound, timbre, pitch, and
duration. The completed piece emerges from the information
thus preserved on the page. Bach is dead, but Bach's work lives;
it exists, it persists, because he wrote it down. From the start, in
fact, composers made a point of recognizing the double identity
their written music possessed as both score and sound. Baude
Cordier's tune completes its circle in the ear and on the page; it
is a round both seen and heard.

For most of music history, the fact that music can be described
by a fairly compact set of information-rich symbols has not altered our broad experience of it. Someone has always had to
translate the mute notes of a score into some action, a movement
of a bow, the plucking of a string, the striking of a key, or the
blowing of a horn. Such actions generate particular responses
from the instrument—it produces the desired sound, then the
next, and the next, with all the nuance and particular effects that
the performer brings to the score, until we recognize that we are
hearing Yo Yo Ma's virtuoso rendition of a Bach suite, the un-

mistakable sounds of medieval polyphony, or a classic recording by John Coltrane. But while such performances derive their individuality from the use of musical gestures—accents, attacks, rhythmic variations, vibrato, and so on—not necessarily recorded in the score, some musical artisans seized on the idea of music as information from a surprisingly early date. Such men realized that they could build machines that could read the abstract symbols of musical notation. The machines they made could then automate a kind of musical performance, to make music directly from the data.

The basic mechanism of an automatic musical instrument, in fact, dates back almost to the dawn of written music, to the eighth century. There are hints (but no instruments) that suggest that even in the midst of the darkest Dark Ages, someone figured out that by sticking pins into a rotating wooden barrel one could trigger keys or pluck strings in a set order. The oldest working barrel organ to survive was built in 1502, and the catalogue of Henry VIII's instruments compiled at his death included a listing for "An Insttrumente that goeth with a whele without playinge uppon." By the seventeenth century, if not before, such instruments had been fully automated, driven by a clockwork mechanism. The instrument's owner would wind a spring, which turned a wooden barrel that was inscribed with a grid on which a tune could be laid out. The spring transmitted its force to the barrel through a gear train, and some of the more advanced versions of automated instruments—spinets, automatic harps, and even flutes—were controlled not just by clockwork but by clocks themselves, spring-driven timepieces.

The barrels serve as a form of musical notation in the round. The circumference of the barrel dictates the time of a musical composition, just as the length of a piece of paper, measured out in bars, fixes the duration of a conventionally written score. The location of a pin up or down the barrel fixed the pitch of a note, while a number of different systems evolved to set the time value for each individual note. Longer pins could correspond to longer notes; a larger gap between pins or a closely packed sequence of pins could also generate a sustained tone depending on the precise design of the string mechanism to be plucked or the key to be depressed. With careful planning and a system for registering the position of the barrel in its crank-and-gear system, a skilled instrument builder could cram a number of tunes onto the same cylinder, six or more. The result could be impressively long per-

A seventeenth-century drawing of a water-powered barrel organ.

formances: automatic organs manufactured early in Queen Victoria's reign could mount three barrels at once, which allowed them to produce extensive chunks of music, note-for-note perfect when compared with the original score. Nonetheless, barrels were relatively difficult to reproduce, making the "publishing" of music for automatic instruments a laborious business, and the barrels themselves could not hold all that much data, which meant that accumulating a library of music had to be a slow and expensive process.

The fundamental difficulty was that, in the anachronistic language of computers, all barrel organs were what would be known now as particularly inefficient read-only devices; they could transcribe notation written in pins into music heard in sound, but they could not write new tunes to their cylinders. The next stage in the automation of music required two separate breakthroughs: first, a much denser, much more efficient data storage system had to be invented; second, a method had to be devised to write music onto such a system in such a way that music could be recorded as well as played back with relative ease. The solutions for both turned on the same late-eighteenth-century invention, grandly named melography—the craft of writing music in the form of holes punched in a paper tape.

The underlying idea originated outside of music. Between 1725 and 1804 a series of French inventors attempting to automate the process of weaving complex patterns into silk worked out a technique that allowed the machinery of the loom to replace the work of the weaver's assistant, or drawboy. The drawboy's task required him to select and "draw" into the pattern the right sets of colored threads to be woven together with the main, background color or colors threaded onto the loom frame. Patterns had to be memorized, and the draw threads strung for each design to be woven—the first a difficult task, and the second a slow one, with changes in patterns consuming two weeks or more of labor and costly downtime for the loom. Despite the name, incidentally, drawboys were often women, up to three per loom in the center of French silk making at Lyons. Speeding up the whole weaving process and replacing that crowd of (however ill paid) expensive laborers became essential as the Lyonnaise weavers confronted ever-increasing competition from cheap Indian and Chinese silks. The first step, taken in 1725, encoded the selection of the draw threads in holes punched in a continuous loop of paper. Each row of holes selected a given set of needles attached to the draw threads. Those threads would then be lowered into the weaving plane by a single assistant, pressing on a lever or a pedal. The device contained the essential, spectacular discovery that machines could be controlled by a series of simple instructions on paper—but it did not actually work all that well. The patterns had to be simple and repetitive enough to be contained on a single loop (like the tunes short enough to fit on a convenient-sized barrel).

The next advance generalized the original idea. A master weaver named Falcon replaced the loop of paper with a stack of stiff punched cards, the direct ancestor of the punch card that was omnipresent at the dawn of the computer age. The cards still had to be fitted to the loom by an assistant, but they multiplied enormously the amount of pattern information that could be stored and read, dramatically speeding up the operation of a draw loom. Each card recorded a single set of choices of threads and could be changed with each flight of the weaver's shuttle to create a pattern as complex and varied as could be imagined. The entire pattern could be changed simply by selecting a different set of cards rather than resetting the loom as a whole. The first loom thus equipped was made in 1739, but the machine as built (unlike the principle) proved difficult to apply to the variety of looms in use at the time. The development of a fully automated loom came only in 1804 when Joseph Marie Jacquard combined the punch-card system of data storage with a mechanical device that could advance and register card after card accurately on a loom. Jacquard's synthesis of the software and hardware of loom control was so complete that the fundamental design has not changed since.

The riots of displaced weaver's assistants had no effect. Within two decades Jacquard looms dominated the silk trade, traveling from France to Italy and beyond. With them spread the idea behind their design. Apparently indefinitely complicated outcomes—like pictures woven into cloth—could be represented in a compact, machine-readable form. Charles Babbage recognized

A portrait of Joseph Marie Jacquard on silk woven on a Jacquard loom.

that such cards could serve as a source of data for his proposed "engines," steam-driven mechanical computers, designed but never completely built in Victoria's England. More prosaically, plenty of musical instrument builders recognized that punch cards were a whole lot easier to work with than precision-crafted barrels stuck with pins. Instead of fixing the length or spacing of a nail in a cylinder, changing the size of a hole cut into paper could control the duration of a note; location on the paper could (as before) fix the pitch; and, most signficantly, the use of stacks of paper cards meant that automatic musical instruments could become universal in this sense: instead of being restricted only to the music hammered into its one or few barrels, a street organ could now sound any tune, as long as it had been encoded upon an interchangeable stack of cards.

Card-driven street organs still exist. Often they are manned by teams of two. One man will stand out front of the instrument, bowing and swaying to the music while cajoling passers-by into making their contributions. Meanwhile, the other minstrel half hides behind the chest and pipes of the instrument, cranking away on the mechanism that both supplies air to the organ and drives card after card through the assembly that feels for holes and strikes the correct keys. The drama enacted by the front man turns on the wonder of the instrument: Give money, he demands, for the gift of having heard music you cannot make yourself, that we conjure without effort out of our magical, mechanical box. Today, of course, for "magical" we read "quaint." But the mummery of the street busker recalls what has been forgotten in this age of ubiquitous disembodied music. With the voice of a singer or the hands of a performer, music had since time began been something that was made, a human experience created by one or many on the spot, to be consumed on that same spot by the crowd or the solitary listener. The automatic organ, though, carried with it far more than just the tunes it played. On one side, such instruments were travelers in time and space, bringing to their audiences music made long before and far away. On another, they revealed a glimpse of an unsettling thought about the nature of experience: music, what the audience heard, emerged directly from a stream of information, data in a form that could just as easily generate a picture of a bird of paradise woven into a piece of silk as it could produce a two-voice fugue on the pipes of a hand-cranked organ.

One consequence of the invention of this kind of universal

code was that data written in this way could be published, readily reproduced—far more easily, in fact, than conventional musical notation. Music scores had always been cumbersome, and, in fact, until the creation of music-writing computer software in the 1980s, music typesetting was so expensive that one of the great roadblocks slowing the performance and dissemination of works was the tedious pace and cost of copying parts by hand. By contrast, punched cards, and later punched paper rolls, were relatively simple to manufacture in quantity. Instead of complex, varied, and difficult to proofread musical symbols, the compositor had only to set up a machine to punch holes of differing lengths. Once the master version had been made, it could then serve as the template for any number of copies, run off on a hole-punching press.

More important, though, the punch cards and paper rolls that permitted such seemingly trivial gains as expanding the street musicians' repertoire contained, crucially, much more than just music, scores written in a peculiar code. Conventional notation holds a record of musical events in an abstract form, which an individual musician had then to interpret and reproduce as best she could. But the information stored on perforated music rolls expanded to include representations of actions, the events that translate a static written score into a performance. Both systems recognized that the pitch known as middle C could be thought of as an abstract datum, no different in kind from the datum that describes D, or the color red in a bolt of cloth. But the perforated roll could also record the force of the downstroke of the performer's right forefinger—the sudden accent or muting of a note—as one more bit within an information stream.

This new, more general concept of information that included both musical data and musical expression made possible a revolution in the way people began to encounter music. The pneumatic, paper-roll–controlled player piano was perfected and patented between 1880 and 1900. By 1930, they were everywhere; two and a half million had been sold in the United States alone. In some sense, all earlier automated instruments had been curiosities. Like the street organ with its costumed beggar-attendants, the older versions were remarkable not so much for the music they made but for the fact that they made anything resembling music at all. The player piano was different. The best were capable of an unprecedented subtlety and nuance of expression; they were the first automated instruments that could plausibly justify the claim that they produced not just tunes but music that could be called art. Long before the gramophone achieved similar quality, the player piano lifted musical performance out of public spaces, concert halls, and carried it into private homes. By harnessing music as information, the player piano transformed information into power: the power to hear the music you wanted, when you wanted to, where you wished to listen to it.

The instrument achieved its results through the successful union of two of the highest technologies of the Victorian era: the Jacquard-based system of programmed machine control managed the player piano's extraordinarily sophisticated network of pneumatic valves and motors—the air-driven devices that actually struck the notes to be played. As Arthur Ord-Hume tells it in *Pianola*, the best single-volume history of the instrument, the precise designs varied with different makers, but all player pianos exploited the possibilities created by setting up a difference in air pressure inside the instrument compared with the ordinary pressure of the atmosphere outside. A player piano sounds a note when a pneumatic valve opens, lifting a rod connected to an assembly that transmits the original valve motion to the point at which it can trip a hammer and strike a string. The process begins with the action of a pump driven by the treadles or foot pedals trod continuously by the player piano's player. The pump reverses the action of an organ's bellows. Instead of forcing air into a windchest, increasing the compressing air to be released by flowing through an organ's pipes, the player piano's bellows suck air out of a chamber, reducing the air pressure inside the instrument by creating a (very) partial vacuum. The valve and the pneumatic linkages connected to the piano hammers for all eighty-eight notes of the piano are tied into the vacuum chamber; the whole system forms a zone of low pressure into which the atmosphere outside the instrument seeks to gain access. When a hole in the paper roll being played reaches the data input device, called a tracker bar, it uncovers a tube that leads from the tracker bar to the valve chamber. The rush of atmospheric air down the tube forces the valve upward, which trips the first of several levers that lead to the hammer and the note.

So far, the pneumatics simply connect the data—the piano roll—to the physical mechanism of sound production, offering no obvious advantage over ordinary, purely mechanical linkages. But the pneumatic system could perform one function simpler con-

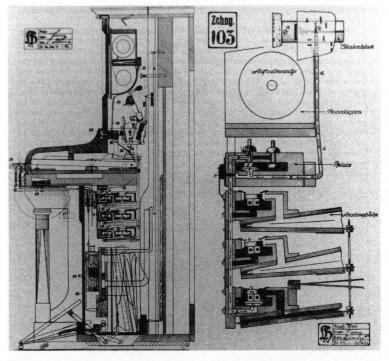

Diagram of a player piano; the detail on the right shows the pneumatic stack—the air pathway that governs the action of the piano.

trols could not: by varying the amount of air allowed into the system, pneumatically driven player pianos could generate a theoretically infinite amount of variation in the notes to be struck, changing volume, attack, and accent just as a human performer would. The control system ultimately adopted most widely was invented around 1900 by a Boston-based inventor named James William Crooks, who sold it to the Aeolian company, which further developed the idea and patented it under the name Themodist in 1910. Themodist accents are coded into an additional double row of perforations along the side of the piano roll. As a piece proceeds toward the note to be singled out, the Themodist perforation allows a little extra air into the pneumatic stack, increasing the difference in pressure outside and inside the piano, forcing the valve of the selected note open faster, driving the levers harder, and striking the hammer onto the string more quickly—thus producing a louder sound, a more sharply attacked note to stand out against the background of the accompanying music.

Using the most sophisticated form of variable pneumatic control, several player-piano companies began as early as 1904 to manufacture what were called reproducing pianos. Reproducing pianos could generate up to sixteen different shades of loudness, and some could vary the volume they produced on a continuous, sliding scale. The best of them were Ampico actions; mounted on various makers' conventional pianos they offered different settings for playing style—subdued, normal, brilliant—and responded fast enough to keep up with the most demanding score encoded onto a piano roll. By the end of the First World War, Ampico models and others had become accurate enough to attract (for handsome fees in some cases, to be sure) the endorsements of the leading pianists of the day.

More important, such pianos enabled those pianists to record their work with far greater fidelity than could the gramophone. Recording pianos contained two sets of electric contacts, one beneath the keys and one near the point at which the hammers hit the strings. The closing of the two contacts would transmit pulses to electromagnets above specially treated recording paper, creating a signal that encoded the pitch, duration, and force with which the string was struck. The electromagnets would then punch a hole or trace a mark to create the master roll for a player-piano performance. The recording apparatus could punch 4,000 holes per minute—enough to store all the notes even the fastest virtuoso could pound on a piano keyboard. By contrast, gramophones in the 1920s could reproduce only a fraction of the sound produced by a piano, losing both complexity and the richness of the tone of the instrument. Even worse, they could play back only four minutes or so of music per side of a 78-rpm record. The pianist Ferruccio Busoni complained in a letter to his wife in 1919 that recording for the gramophone "is stupid and a strain." He explained that, for example, "they want the Faust Waltz (which lasts a good ten minutes), but it was only to take four minutes. That meant quickly cutting, patching and improvising . . . not letting oneself go for fear of inaccuracies and being conscious the whole time that every note was going to be there for eternity; how can there be any question of inspiration, freedom, swing or poetry?" Making a gramophone record was a hit-or-miss game—either the performance worked or it did not. Recording a piano roll, though, was a far different proposition. Because the rolls stored data, not sound (as in the grooves of a gramophone platter), the artist making the recording gained extraordinary control over the outcome. The pianist using a recording version of a reproducing piano could play as long as he liked, play with all the expression he chose, and make mistakes.

And at this point the difference between the music itself (the performance as it had traditionally occurred) and the transformation of a performance into data, into abstract information, became crucial. Because a piano roll stored not sounds fixed for eternity but a set of parameters that described the sound, performance itself remained malleable; the information could be processed, the holes recut, until the performer was satisfied. With the master roll in hand, the performer and the recording engineers could go in, mark any wrong notes or changes in expression, punch them in and play them back as desired. The result, in theory, was the perfect recording, the exact intention of the artist.

That prize caught the attention of the leading musicians of the day. George Gershwin recorded for the instrument; so did Richard Strauss, Alexander Scriabin, Paderewski, Schnabel, Rachmaninoff, Debussy, Fauré, and many others. Stravinsky collaborated with a French music publisher to produce piano transcriptions of *Petrushka, The Firebird, The Rite of Spring* and other works—some fifty rolls of music in all. The advertising claim made by most reproducing-piano manufacturers was that owning one of their instruments allowed the listener to hear Stravinsky playing right in her own living room. The reality was that the reproduction of the sound was no better than the condition of the piano playing back a Stravinsky roll, within the acoustics of the room in which the music was being heard. But in the best conditions—hearing the music played back on the same piano it was recorded on, in the same studio—Rachmaninoff was sufficiently impressed to state: "Gentlemen, I have just heard myself play." Percy Grainger, a student of the gramophone-hater Busoni, went further, announcing that the piano roll portrayed him as greater than he was, performing not as he did, but as he "would like to play."

Dangerous words, those—ones that suggest the implications of constructing a musical experience out of musical information rather than out of an immediate action. To imagine being heard as one dreams of playing, with the aid of a machine that reproduces sound represented as data on a strip of paper, suggested a basic change in how music comes to be. The musician had been a kind of craftsman, manufacturing a product within the traditional constraints of the craft shop. Just as Stradivari made violins one by one, by hand, so someone playing a Stradivari instrument would stand for half an hour, producing sounds to be heard just that once, only by those in the same room. By contrast, the player piano worked by breaking down a complex phenomenon into its simplest components, its bits of information. Information—a book, a score, a roll with holes, bits on a magnetic disk—travels easily, far more so than the things such information represents. Information in the sense of the data stored on a piano roll is a quantity, a commodity, one that can be mass produced and broadly distributed. The impact was tremendous. We now live so thoroughly surrounded by music, so inundated with our options, the recorded performances we can hear at any time, in any order, exactly as we choose, that it is virtually impossible to imagine what it was like to be limited to only the music made for the purpose, for the moment. But it was the player piano, those two million and more instruments spreading out across America, that bridged the gap between the old era and our own, transforming forever the relationship of the audience to its music. For the first time, those who listened to music gained a substantial measure of control over what music they would hear, when and where.

And just as the production of music as information altered the role of the audience, it shifted the ground beneath the composer as well. Information-controlled instruments freed composers and their music from the constraint of the performer and the limits of his abilities—the fact that Percy Grainger had glimpsed. The obvious next step was to produce music that machines could make that would be beyond the abilities of any mere, ten-fingered mortal. The most spectacular piece composed with a player piano specifically in mind was written as late as 1977 by the Netherlands' Jan van Dijk. He composed a piece for player piano and orchestra, performed with a piano roll prepared by Lucius Voorhorst. Before the piece could be heard, Voorhorst had to program Van Dijk's score onto a piano roll, positioning each of the approximately twenty thousand perforations that represented an idea of sound, nine minutes of solo player piano. The piece thus encoded included passages with as many as sixteen notes playing at the same time, a musical event that no human performer could produce.

In other words, the creation of mechanical devices that could translate data into sound expanded the universe a composer could explore. Making music out of information gave the composer more stuff to make music out of, more noises, more choices, and more decisions. To return at last to Michael Hawley's ultimate player piano: his computer-controlled Bösendorfer represents the final form of what was essentially the nineteenth-century discovery of the connection between information and experience. Hawley uses his piano mostly as a research tool to investigate, among other issues, what physical aspects of performance distinguished one pianist from another. The Bösendorfer, together with Hawley's custom software, reproduces the actual physical events that a pianist uses to produce nuance. The system then writes those events as digital information—which Hawley can then use, just like any other piece of data, to shape any sequence of notes he commands his machine to play. Some of what he does is more straightforward. Just as piano-roll makers in the twenties encoded four-hand music, Hawley can take the data off a piano roll and add instructions, so that the result is not what the pianist played, but what the pianist might have done given the extra limbs. One of his showpiece demonstrations is the transcription of a Rachmaninoff roll, marked by a spectacular glissando—a roar up and down the keyboard in a cascade of notes. The original version is impressive enough, fast, flowing, and, as corrected on the piano roll, absolutely accurate. Hawley's version, though, doubles the line, multiplying the sound to create a note-perfect rise and fall of incredible density. It sounds exuberant precisely because the result is audibly other than, more than, what a human being could have played.

And that creates a problem. In practice, in the twenties, at the absolute height of the player piano's popularity, very little such experimentation took place. Manufacturers were too busy shipping as many popular tunes as they could publish. But nonetheless, the sequence of technological and musical ideas that led from the home player piano to Hawley's nearly unique great beast seemed obvious to at least some musical thinkers in the twenties and thirties—an obvious road to travel, and a disturbing one. Béla Bartók, speaking in 1937, delivered a harsh warning against succumbing to the lure of what he called mechanical music. Bartók noted that Stravinsky's motive for writing a piece for the player piano was to eliminate the performer's chance to interpose his personality between the composer and the ultimate shape of the music to be heard.

Bartók added dryly at this point: "Whether or not this principle is correct is an entirely different matter." The bulk of his lecture then treated a range of issues raised by gramophone recordings and radio broadcasts, but at the end of his talk, the composer returned to the underlying issue of constructing music out of stored data. Mechanical music fails as music, he argued, because it eliminates one critical feature of live performance. "That which lives changes from moment to moment," Bartók said, while "music recorded by machines hardens into something stationary ... mechanized music cannot be a substitute for live music; just as a photograph, no matter how artistic, cannot be a substitute for a painting." The ultimate threat, Bartók warned, was that mechanical music could "flood the world to the detriment of live music, just as manufactured products have done to the detriment of handicrafts," and, he concluded, "may God protect our offspring from this plague."

One may judge for oneself how Bartók's prayer has been answered—but the underlying issue he raised remains open. The essence of what offended him about constructing musical experience out of musical data was the shift in where the essence of music lay. In mechanical form, the music is fixed and eternal; it lies within the sequence of holes on the page, grooves on the record. To Bartók, such music was anathema. By contrast, Bartók held that the the information required to make music, the composer's intention as represented by the score, contains just the description of what the music might be; the actual music exists only as it is expressed in sound by people, for people. But while Bartók said what many felt, there were those who regarded such claims as romantic nonsense. Arnold Schoenberg ended up in this one arena more or less on the same side as his rival Stravinsky. He conceded that the idea of a perfect, single interpretation of a work was ridiculous, and he dismissed Stravinsky's desire to express the composer's intention perfectly, forever, by means of a mechanized performance (though he longed for just one musician "with the will and the ability to discover from the score what is true and eternally constant"). But with Stravinsky and against Bartók he recognized that the mechanization of the production of musical sound gave him the chance to expand both his palette and his control. Welcoming the new automated instruments, anticipating objections like those Bartók later made, Schoenberg argued in 1926: "Ensuring the production of sounds and their correct relationship to each other, freeing them from the hazards of a primitive, unreliable and unwilling sound-producer—to that degree the use of all mechanical musical instruments could be of the greatest advantage."

Schoenberg dismissed those who complained about the loss of

the human soul of music. "It is sentimental," he wrote, "to wail about mechanization and unthinkingly to believe that spirit, so far as it is present, is driven out by mechanism." He believed the loss of the freedom, of the interpretative variability, of live performance was more than offset by the gain in accuracy made possible by what he called "the mechanization of music." The objections of those frightened of such music or of the loss of the role of human musician provide "no reason," Schoenberg concluded, "to begrudge them and us better instruments."

With hindsight, it is clear who won the argument, whether right or wrong: automated instruments of all types abound, not to mention recorded music. But despite his defense of ever more advanced instrumentation, Schoenberg placed a limit on his indulgence of musical license. The free flow of musical information from the composer's brain through some system of automatic control to an output device like a player piano was good, for it freed the composer from the confines of human frailty. But on some level Schoenberg lost his nerve, though he had conceived of music liberated from the conventions of harmony that dated back to Bach and the music made with the well-tempered scale. The idea that pieces are "in" a key dominated Western music from before Bach's time, with compositions organized around the harmonic relationships that could be built out of the scale that began at D, for example. Schoenberg rejected the idea that one note, one tonality, ought to have primacy in a piece and developed a system that treated every pitch of the twelve-note octave entirely impartially. Beginning in 1923 he proposed that each piece should be built around individual sets of all twelve tones in the scale, with the composer's goal being to use all eleven of the other notes before repeating the twelfth, thus ensuring that no one note stands out in the mix.

Schoenberg succeeded, in his own music at least, in breaking apart the confining rules of Western harmony; his twelve-tone music has a geometric kind of order, a balance of spaces and distances between notes that can be seen on the page as well as (or even more easily than) being heard. But having exploded one set of rules that governed musical experience, he faltered. Even as he seized for himself an unprecedented degree of freedom in how he used his sounds, he began to codify the regulations that would govern matters in his new musical state. It seems as if, having broken loose of the old order, he found himself suddenly afraid that the entire idea of order might have been swept away, and sought, as successful revolutionaries often do, to reassert a rigorous control over the situation. "The main advantage of this method of composing with twelve tones," he wrote in 1941, long after he had first conceived of the idea, "is its unifying effect." Unity was the highest virtue, above and beyond the musical effect being sought. In a 1946 postscript to his 1941 essay, he complained that Alban Berg had slipped into error by mixing old-fashioned tonal, harmonic music in with his orthodox, pure, atonal work. Berg's excuse was that his operas needed to use tonal material for dramatic effect. Schoenberg acidly commented: "Though he was right as a composer, he was wrong theoretically."

It is easy now to poke fun at Schoenberg, transformed almost instantly into a twelve-tone commissar, from radical to conservative—and it's just as easy to see Bartók as a kind of Luddite, railing away too late against the development of a technology that has, after all, made music available to anyone. But both men looked into the face of an idea of music, of experience in general, that they clearly recognized as new—and between them they identified two of the greatest dangers that would accompany such an idea of the world. Bartók understood that the reduction of music (of experience) to its simplest, interchangeable elements posed the risk of drowning out any given musical event in the rush of noise. One signal built out of bits of information could be easily lost, he saw, in a sea of seemingly identical data, the cacophony of all the information available. Schoenberg, meanwhile, worried less about picking out the signal from the noise—but trembled at the thought of being unable to interpret any message, unless some rules, at least, remained in place. Without answers to these two questions—how to isolate the data that matters for a given enquiry, and how to understand what that data might mean once it has been found—the danger lay that music (and perhaps the world) would become unintelligible, meaningless.

That was and remains a real danger—there is plenty of music being made now that cannot be heard for the clutter, or that sounds simply arbitrary, opaque, and dull. But even as Schoenberg and Bartók confronted their musical choices, the next change in the technology of making music was overtaking them. The inventions they confronted, the player piano and the gramophone, proved that information made or found in one place could be transported to another—but that was all. The player piano decoded what was already there, but it did not alter it, process it, or make it. But once it became clear that music could be manipulated as data, the next step became obvious, leading to instruments that would not only use data to make sound but would create, modify, manufacture, information on their own, making sound at will. Such instruments now contain an entire universe of musical choices in a box—and with such choices, all the attendant dangers and possibilities.

The first steps toward the creation of truly flexible musical instruments were in some ways steps backward. The early electronic devices transformed current into vibrations, sounds, rather than relying on the mechanical energy required to make noise with a violin, a piano, or an organ—but the new instruments, like the old, simply translated energy of whatever source into one particular type of sound. One of the most famous was invented in the early 1920s by a Russian cellist and radio engineer named Lev Termin, or Leon Thérémin, as he was known in the West. The device, called the theremin, had two antennae, with electric current flowing through them, connected to an amplifier and a speaker. The performer played the instrument by moving his hands near the two antennae, altering the electromagnetic fields formed by the antenna current and thus determining the pitch and volume of the sound produced. (The theremin, that is, employed something like the effect that can drive television viewers crazy when they try to improve the image on the screen only to find that the picture decomposes as they pull their hands away from the antenna.) The device made its concert debut in 1924 with the Leningrad Philharmonic. The instrument grew rapidly in popularity, partly as a result of demonstrations during which Thérémin would sculpt the air to produce Schubert's "Ave Maria" or Saint-Saëns's "The Swan." RCA decided to manufacture theremins in 1929, and composers began to make use of the device in increasing numbers. Charles Ives wrote for the instrument, as did Aaron Copland and the player-piano devotee Percy Grainger. Theremin music could be heard as well on the sound tracks to movies like *Spellbound, The Lost Weekend,* and *Alice in Wonderland*—and even in the opening notes of the Beach Boys' hit "Good Vibrations."

The theremin had a ghostly, unearthly sound, a nearly pure fundamental tone with few overtones, which, combined with its ability to slide smoothly and continuously up the full range of tones audible to the human ear, made it an instrument of choice for conveying a sense of space, of surreality. It was limited, of

Lev Termin, playing his theremin circa 1927.

course, just like the other major pre-World War II electronic instruments, in that while it made different noises, new ones compared with those of preexisting instruments, it still offered just one more such sound for a composer to use. Nonetheless, the theremin suggested to the most daring of musicians what they could expect from a musical instrument, what kinds of new music they might be able to create. The French composer Edgar Varèse had complained as far back as 1916 that "we need new instruments very badly." The theremin answered part of his need, and he composed a piece that used the instrument in 1934. Its ability to generate any pitch, unconstrained by the architecture of the octave and the scale, attracted him, and then spurred him on to new heights of desire. In 1939 he wrote of what he wanted next from the technology of sound production: "Liberation from the arbitrary, paralyzing tempered system . . . any number of cycles . . . consequently the formation of any desired scale; unsuspected range in low and high registers, new harmonic splendors . . . new dynamics far beyond the present human pow-er[ed] orchestra; a sense of sound projection in space by means of the emission of sound in any part or in as many parts of the hall as may be required"—in other words, total control over and total freedom to choose the pitch, timbre, loudness, duration, and location with respect to the listener for any note, any sound that Varèse or anyone else wished to employ.

In 1939 such desires were so many fantasies, dreams—but the technology being sought was not that long in coming. Between the theremin and its various cousins, and the modern inventions that substantially met Varèse's demands, lay one major intermediate technological step. The early electronic instruments that became truly commercially successful were those that imitated conventional designs far more closely than did the theremin. The Hammond organ, invented in 1935, behaves like an organ with a distinctive sound—one to which 1950s-era rock-and-roll owes an enormous debt. The electric guitar, invented after World War II by Les Paul, achieved the same end. An electric guitar creates a sound by using an electronic coil that sets up an electromagnetic field in response to the vibration of a wire string, which generates a current that can be amplified and turned into sound with a distinct pitch and timbre. But while the means of sound production is new, the instrument otherwise acts just like an ordinary acoustic guitar, sounding a note when someone strums a string. Although rock-and-roll would be lost without it, the electric guitar did not liberate music from the constraints Varèse sought to escape.

Nonetheless, the electronics that lay at the heart of all these new instruments did point toward a source of sound that could meet Varèse's goal. The creation of a technology that could take an electronic signal and produce a sound with a defined pitch and color suggested the possibility of what came to be called sound synthesis: the manufacture of tones created by conceiving of and manipulating electronic pulses, electronic information. The idea of a musical synthesizer as a universal instrument, designed to produce a wide variety of sounds, emerged in the 1950s. Harry Olson, an RCA engineer, defined what such an instrument ought to achieve: "The synthesizer," he wrote, had to generate "any kind of sound imagined. Then if a person can imagine a hit, the synthesizer will facilitate the production of the hit."

The synthesizer Olson and his colleague Herbert Belar developed fell far short of that goal and never made it to market (though it was installed in the Columbia-Princeton Electronic Music Center). It possessed too few signal sources, electronic oscillators that could produce waves to be translated into sound, and it contained too small a range of the different electronic circuits that could shape such signals to produce a wide range of sounds. It was utterly unable to emulate even remotely the complex sounds produced by Olson's target instruments, the clarinet and the oboe. It did, though, contain one important feature: it updated the player-piano control system, using a punched-paper tape that contained the program for the whole series of actions needed to assemble sound within the machine. Nonetheless, in practice sound synthesis evolved without the aid of true synthesizers (that is, machines capable of constructing a wide variety of complicated sounds from scratch) for more than a decade. Work on the problem accelerated first and fastest after World War II in Paris, where the composer and engineer Pierre Schaeffer developed what he called *musique concrète*. A *concrète* composition began with recorded "real" sounds—of a train in a freight yard, occasionally orchestral instruments, human breaths, footsteps, a whistled tune, and so on. The tapes of those sounds could then be manipulated by a variety of operations including those that would alter the frequency, amplitude, even the order of the electronic signals on the original recorded sound source. Following a line of research that pushed closer toward the idea of fabricating from scratch any sound desired, the Electronische Musik (Electronic Music) studio in Cologne used simple electronic oscillators to generate electromagnetic waves that could be translated into pure sounds. The studio possessed an oscillator that generated smooth sine waves, and others that produced square and sawtooth forms. The frequency of the waves determined the pitch of the note being made, while the shape affected the timbre. Using a variety of amplifiers, filters, and analyzers to manipulate taped material, the Cologne group could build ever more complex sounds out of the initial simple wave forms, altering the attack, decay, and color of each note.

Such techniques did offer part of what Varèse had sought: there was no need to stay within the confines of the scale, and the studio continuously developed new sounds previously unavailable to composers. Among the important early works produced by the studio was Karlheinz Stockhausen's *Kontakte*, written for tape-recorded electronic sounds, piano, and percussion in 1959–60. For Stockhausen, the piece was the culmination of a series of experiments in the shaping of the individual elements that could be generated electronically to construct an increasingly compli-

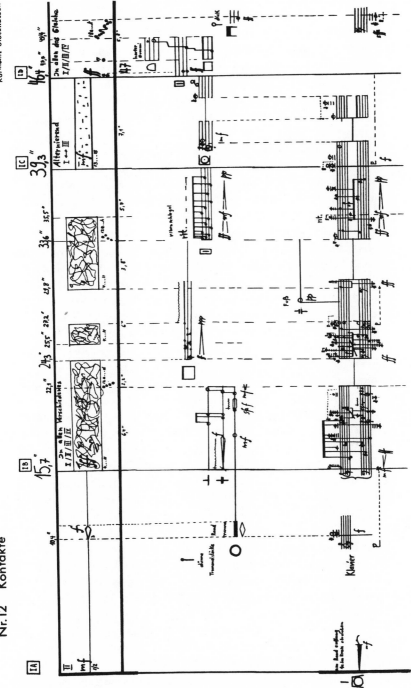

First page of Karlheinz Stockhausen's score for *Kontakte*.

cated and rich set of sounds. *Kontakte* was, he wrote, concerned with (among other things) "the contacts ... between different forms and speeds in different layers" of sound in the piece. He continued: "Many different sounds in *Kontakte* have been composed by determining specific rhythms and speeding them up several hundred or a thousand times or more, thereby obtaining distinctive timbres." Working out ways to build sounds for use in the piece led Stockhausen to turn the process on its head, to make music by disassembling sound. He wrote: "If we can compose these sounds, in the sense of the Latin *componere*, meaning put together, then naturally we can also think in terms ... of the 'decomposition' of a sound. ... The original sound [twenty-two minutes into *Kontakte*] is literally taken apart into its six components, and each component is decomposing before our ears into its individual rhythm of pulses."

Kontakte remains an astonishing piece to listen to, exciting, dramatic, dense, and enormously complicated. It has a rough, harsh-edged feel to ears conditioned by more than a decade of the smooth and seamless qualities of digitally synthesized music, which adds to the impact of the older work, heard fresh. But what makes *Kontakte* stand out most now are the ideas about music—both past and future—that it explores. A fair distance lies between the player piano and the machines of the Cologne Electronic Music studio, but there is a link that connects the two. Both technologies broke the connection between the action of the musician and the production of a musical sound. Each interposed a stage in that process during which the music exists solely as a concept, as a message, information written down (holes cut into a roll of paper, settings for a series of electronic circuits) to be transmitted from the place where it was written to the place where it can be read. *Kontakte* served Stockhausen as a kind of test bed in his attempt to isolate and identify the elements— data, electronic signals—that could be assembled into sounds. The technology at hand was crude: his decomposed note was built out of only six elements, after all, a trivial handful compared with the complex sound production that takes place within as common an instrument as the cello. But despite the limitations of his tools, Stockhausen managed to realize the critical advance. Once he had found his elements, he demonstrated that isolating sound as data enabled him to derive an enormously varied new source of musical effects. By constructing his sounds piece by piece, datum by datum, he gained the ability to manipulate at will the music to be heard, exposing to the ear combinations of sound never before encountered, never before imagined.

Stockhausen's tools, though, were not merely crude: they were also cumbersome, slow, difficult to master, expensive to use, and still enormously limited in what they could do. Each sound had to be constructed more or less by hand, by selecting and tuning control knobs, cutting and splicing tape recordings to build up the entire catalogue of desired characteristics. The full potential of music so constructed from scratch could not be tested until someone had built the instrument that could truly automate the job—one that could, as Olson had declared, provide the musician with any sound that could be thought of. The next to last step toward that goal was taken in 1964, when an engineer who got his start in a factory that made theremins built the first of what would become the legendary Moog synthesizers. The synthesizer Robert Moog built employed transistors in place of the older, clumsier (and slower) vacuum tubes—and it incorporated the fundamental innovation that formed the heart of virtually all subsequent synthesizer designs through the 1970s. Older systems, both the RCA synthesizer and the complicated chains of equipment stitched together in the early electronic music studios, worked by creating an electronic signal in one place, transmitting it to the next circuit or device, modifying it there, and so on, in an inflexible, linear chain of operations. The small size and flexibility of transistors opened up a new possibility, involving the use of a technique called voltage control.

Voltage-control systems work by hooking up a source of both positive and negative voltages to a circuit, whose output can be controlled by a device called a potentiometer. Adding or subtracting voltages can determine both the frequency with which a transistor drives an oscillation, determining the pitch of a note, and the amplitude, setting the volume. So far, this is nothing new—the principle involved is little different from that which governed the theremin. But critically, hooked up to transistorized systems, the voltage controls could manage the entire collection of sound-making or modifying circuits, passing the same information—the voltages—down a series of connectors hooked to each device. And finally, they allowed the musician to add new information, new voltages, to circuits down the line, using a new power source with its own, separate, user-determined setting. What this meant in practice was that for the first time it became possible to add large numbers of signals together in a flexible, continuously changing manner, building an increasingly large collection of sounds, all under the complete control of whoever was minding the machine. Moog and others constructed synthesizers containing circuitry that could change the "envelope" of the sound wave, giving it different attacks or accents and varieties of decay. They could split signals in two and play with the perception of where the note was coming from. They could alter the timbre of notes by linking circuits together, using the output from one as the control voltage for the next, thus manipulating how much sound went into the fundamental pitch and how much into the overtones that give color to a tone. All this could be controlled at a keyboard, with the keys that function as sliding switches, capable of turning on and off particular voltages.

For all that, voltage-controlled synthesizers (also known as analog synthesizers) did fall short of the ultimate goal. They sounded new; they produced lots of sounds; but they sounded like themselves, a distinctive "electronic" palette of musical expression. They still could not fool someone into imagining an oboe hidden behind the wall of electronic gadgetry—which meant that they did not contain the entire universe of sound that those who made music from Varèse onward increasingly had come to believe ought to be theirs for the taking. (To be fair, many of the pioneers of electronic music disdained the idea of making imitations of perfectly adequate existing instruments. They were after a new sound from the start—but in other quarters the desire to find a universal source of all possible musical sound persisted.) Moog and other brands of synthesizer did command a large and growing market, showing up in popular music, jazz, and avant-garde compositions alike. But by far the most popular synthesized works were such transcribed readings as Walter (later Wendy) Carlos's *Switched On Bach*—more a spectacular curiousity than a work of art.

The analog synthesizer era ended, more or less, around the end of the 1970s, though the Moog sound remains part of the rock-and-roll style. Partly the novelty of the sound wore off—the *Switched On Bach* that had at first sounded witty and new came to seem thin, trite, and shrill. But mostly it was the digital revolution that killed the instruments, an upheaval that rendered almost a century of work on electronic instruments obsolete, almost overnight. Sophisticated analog synthesizers had by the 1970s already begun to incorporate digital features, using devices that altered voltages in steps to serve as the binary, on–off switches that perform logical operations in computers. With their logic circuits, such synthesizers could be programmed to activate different sound-producing devices, or to alter the preset controls for any number of synthesis circuits up and down the chain. But the underlying mechanism remained the same: the sound was created by shaping the flow of current through a series of machines, a system that was distant kin to the idea of using a keyboard to regulate the flow of air through one or more organ pipes. Analog synthesizers were more information rich than the organ, if only because the use of electronic signals rather than the mechanical force of air both compelled and enabled the musician to construct each note to her own specifications. But she was still tied in the end to only those sounds that could be made using the outputs of the specific devices contained within her synthesizer. With the invention of the microprocessor, the transformation of the process of making music anticipated (and feared) by those working between the wars became complete: all sounds that can be conceived can be made, are now available, heaped at the feet of anyone who wishes to make use of them.

Compared with the complex mechanisms that drove the first automatic instruments, or with the intricate circuitry required to produce the early versions of synthesized sounds, digital musical instruments rely on an essentially simple design. A digital synthesizer is essentially a computer with customized features that allow it to treat sounds the way ordinary computers manipulate symbols. In digital synthesis, calculation takes the place of the analog synthesizer's electronic circuits. The process begins with the translation of wave shapes into strings of numbers. Wave forms can be analyzed and disassembled using a variety of mathematical techniques. The most important one is called Fast Fourier Analysis, which breaks down a complex wave into the entire catalogue of its component, simple, S-shaped sine waves. Such curves can be defined numerically—and with that, become objects the computer can process. A digital oscillator can generate the string of numbers that represents a given wave; each mathematical operation performed on that string of numbers generates new numbers until a list of data, of numerical information, has been created that corresponds to the particular characteristics desired for the final sound.

Ultimately, because a computer can manipulate any sequence of numbers, perform any calculation, it can in principle produce any sound at all—which formed the basis for the first experiments in digitial synthesis in the late 1970s. Early versions, like the Synclavier, invented in 1977, constructed every sequence of numbers, every sound, from the ground up, in a system that put a premium on programming skill. The great leap forward came with the invention of sampling techniques that added to the sheer calculating power of the computer the ability to treat the outside world as data. Samplers work by sensing a sound, using a microphone or similar device to convert the mechanical energy in the acoustical wave into an electronic signal. The sampler then measures the strength, frequency, and precise shape of that signal instant after instant, assigning a numerical value to each moment—which the computer can then treat like any string of numbers generated by any means, as simple grist for the calculating mill. Once sampled, a baby's cry, a car horn, Yo Yo Ma's cello, Keith Moon's snare hit, anything, can be reproduced, transposed, reshaped at will. The key to successful sampling is to measure often enough so that what is left out between samples— the information lost in the gaps—creates errors in the representation of the sound that are too small to be detected by human ears. "Often enough" turns out to be in the tens of thousands of times per second—accuracy available within consumer electronics since the mid-1980s.

The last piece required for truly flexible digital syntheses was the construction of a system that can easily transfer digital information from machine to machine, from samplers to synthesizers, from drum machines and keyboards to computers, from anything to anything. The Musical Instrument Digital Interface, known as MIDI, set both hardware and software standards, fixing the circuit design and data transfer rate for musical information. Adopted in 1983, MIDI is now ubiquitous; with it came the completion of the technological tools needed to treat music as information from the composer's first conception up to the point when it can actually be heard as sound. Put another way: the computer, the synthesizer, the sampler, and MIDI add up to a system for representing an entire world of experience, that of music, of sound, as a catalogue of infinitely malleable digital information. What we as human beings encounter as a continuous, "analog" reality—one where the middle C we hear rings out and then decays into silence at the moment Emmanual Ax strikes the key, remaining always and forever a middle C—the computer recognizes simply as a signal that can be altered or let stand, one of any number of signals that can be taken to represent all there is to be heard.

The consequences of constructing music purely out of data exposes to view the broader issue of how we are to construct meaning out of any set of data. Music, that is, once again stands in for the universe as a whole, just as it did when Pythagoras built a cosmos on the scaffolding of the scale. The formal name for the discipline that analyzes these issues is information science. That subject first made its appearance in the form of a highly technical paper called "A Mathematical Theory of Communication," published in 1948 by a visionary mathematician named Claude Shannon. Shannon's research at AT&T's Bell Labs focused on the mathematical definition of information, on the techniques needed to measure the amount of information present in a signal, and on the capacity of any channel carrying information. Information science as Shannon initially conceived it was a specialized inquiry, of particular concern for the design of communications systems and of codes. But of that apparently narrow and specialized work the executive director of Bell Labs, Robert Lucky, wrote: "I know of no greater work of genius in the annals of technological thought."

And in fact, since Shannon's first papers, information science has become one of the fundamentally new fields produced by twentieth-century science, providing tools of analysis for understanding such disparate phenomena as evolution and computer design. The field has remained as Shannon began it, mathematically dense and complex, but good lay introductions include Lucky's *Silicon Dreams* and Jeremy Campbell's more speculative (though somewhat dated) *Grammatical Man*. But to pull one thread out of a complicated weave of ideas, the critical point is that Shannon and his successors established a concept of information as a fundamental substance, a category equivalent to matter or energy. That concept led to the creation of methods that could count up totals of information, check on the particular state of a quantity of information, or could transform one lump of data into another—in the same way that physicists had come up with techniques to measure the amount of matter present, or to determine its form, or to convert it from its status as matter into energy. In the most simpleminded of calculations, we now know that Beethoven's fourth piano concerto (to use one of Lucky's examples) contains 211,000 bits of information by one measure, while all of Hamlet's speeches add up to 70,000 bits. All the while each bit of Beethoven looks exactly like a bit from Hamlet—or a bit of anything else. A hit song as "published" digitally consists of nothing more than a string of ones and zeroes, each the same as the next zero, the next one.

As far as information theory is concerned, in fact, all of experience—not just music—can be imagined as being constructed out of different sequences of fundamentally similar bits of data. Each experience forms a particular signal—some fact, some actual truth about the universe to be mined from the never-ending lode of bits. In this context music is simply the human craft that has moved first and fastest into the process of transforming experience into information. But there is a catch: information science is concerned centrally with measurement, counting up the amount of data around and about and testing it for accuracy. The apparatus of information science built up since Shannon's initial work provides an extremely powerful set of tools to decide whether or not any given signal or sequence of information has made it from one end of the process to another without errors. But that theory does not utter a word about how to judge the meaning of the message. So given all the data we can collect, and all that we can do to extract still more out of our machines, the issue becomes: How are we to make sense of the results?

Put another way: a digital sampling synthesizer can construct a scale composed out of sheep bleats, pitched perfectly, but we still have to decide if the sounds of sheep work in the midst of an orchestra, if what we hear adds up to music. There is no single, absolute answer to the general question being asked: how to determine whether any result in science (or any work of art) is correct (or beautiful). But the experience of attempting to make beautiful music with the unlimited resources of modern digital synthesizers suggests what a partial answer to that question might be.

One of the most advanced versions of information-processing musical instruments is the hypercello, developed by the MIT Media Lab composer Tod Machover and a group of engineers led by Joseph Chung. In the hypercello, an electric cello (the cello analog to an electric guitar) serves as the input device to a stack of machines that process the data created by a performer and generate a variety of different types of responses. As the cellist plays the cello, sensors tell one bank of custom electronics the angle the cellist's bow-hand wrist forms, where the bow is on

Philips advertisement for digital compact cassette, showing a portion of U2's hit "Mysterious Ways" in its digital form.

the cello's strings, how much pressure the cellist is placing on the bow, and where the cellist is fingering the strings on the fingerboard. The special circuitry hooked up to the cellist's sensors then translate that analog data into digital information and transmit the results to the system's main computer, which can then transfer that data as MIDI information to a bank of samplers and sound synthesizers. At the same time, the electronic signal of the tune actually being played gets fed directly into a synthesizer, which can either manipulate that data or pass it through as a (sort of) conventional cello sound.

To execute musical functions, Machover and Chung developed a set of software tools to control the flow of information from a performer to his machines. The bow sets up vibrations on the cello's strings—just as in ordinary versions of the instrument—but at the same time it can act as a conductor's baton, with the position of the tip of the bow, the length of bow drawn across a string per second, the pressure on the bow all controlling musical parameters: the numbers of musical lines called up from the computer, the timbre of the pitches being played, the attack, and so on. Machover's interest in timbres in particular led to the development of a technique designed to permit the performer to add the pure cello sound coming off the strings to sampled or synthesized timbres stored as MIDI data within the computer, allowing the performer to decide precisely how much of his own playing he wanted to hear. Another system, called trigger and control, relies on the computer's ability to follow the score of a piece. The computer can then generate an accompaniment to the cellist's solo, one whose pace and rhythm would be controlled by the performer. As it executes each of these functions, the main computer processes the data it receives, modifying or even creating sounds, depending on the instructions contained within its software. The software here is both part of the instrument, an element of the system that generates a sound, and part of the score, the list of the composer's choices of what sound should be heard when. The system as a whole thus acts as both an instrument to be played by the performer and as an orchestra responding to the performer's commands, as if the cellist were a conductor as well.

So far, Machover has written just one piece for the hypercello, *Begin Again Again . . .* It is, like Stockhausen's *Kontakte*, at once an expression of musical thought and feeling and a report on the current state of the art of the technology of making music. It was premiered by Yo Yo Ma in the summer of 1991 and has been performed a number of times since. At its European premier in Amsterdam's Concertgebouw in the spring of 1993, *Begin Again Again* ... served as the culmination of a week of Ma performances that traveled from the music of Bach performed on a baroque cello through to the Machover work.

The performance had its moments of unintended comedy. The Concertgebouw is one of music's cathedrals, one of the very best concert halls in the world. It preserves the nuance and detail of a performance better than virtually any other major space in the classical music world, so that the audience hears every note, every gesture that a musician has the skill to make. Machover's piece, though, can only be heard through loudspeakers, whose blocky presence, suspended above the audience, burst like strange growths into a room built for live performance. As the last rehearsal began, Chung and his colleagues Andy Hong and Fumi Matsumoto had to test each component of the hypercello. The amplifier, mixing board, and speaker got their workout with the sound of a rap hit, playing at arena volume—probably the first time such music has been heard inside the Concertgebouw's near-sacred main hall. Meanwhile, Ma sat to one side on the stage, playing Machover's score on his Stradivari, bowing hard, visibly exerting himself—and producing no sound that could be heard above the rap, not even by himself.

There was, in other words, an apparent (and partly real) clash of cultures between the technical wizards and the priesthood of classical music. (Martijn Sanders, director of the Concertgebouw, asked at one point in a rehearsal of Machover's piece: "Does it have to be so loud?" Hong looked up from his mixing board when Ma played a short passage from one of Bach's suites on the electric cello and said: "That cello never sounded so good. How does he do it?" As the saying goes, same planet, different worlds.) But such incidental byplay aside, Machover constructed his piece to produce a musical experience "up to the threshold of complexity that people can listen to." Beyond that, he sought to "hone in on what's important in the information, both for performer and audience." There are, of course, any number of ways to communicate something. Mozart, says Machover, used the minimum, elegant and simple ideas and musical figures. But "somehow Mozart's music is refined to the bare essentials, to the point that the smallest changes are meaningful, so that you can never exhaust the resources of the piece."

Machover's concerns, though, as expressed in *Begin Again Again* ... reflect what he sees as current reality: "Music these days," he says, "is like boating down white water on a raft. The raft is some musical structure that you can hold on to, while the white water is all the additional information out there that is always threatening to tip the raft over." The hypercello system allowed Machover to generate that effect, to float the original cello in a maelstrom of additional sounds, but, he argues: "I think that way not just because we have the machines now but because I genuinely see the world that way. There is an incredible density of experience out there, within which we are constantly trying to find the threads of connections."

In *Begin Again Again* ... Machover used one core musical device to establish this image of rafts of meaning tossed in a sea of information, establishing a dialogue between the lead cello part and the accumulating density of sound generated at the cellist's command by the computers and synthesizers to which he was bound. As performed at the Concertgebouw, Ma's line offered a sense of structure, of musical development, while the computer provided the context out of which that structure emerged and within which that sense of order contended. And that piece in that performance worked—which is to say, the audience responded to it, engaged it while it unfolded, and awarded Ma and Machover a standing ovation at its conclusion.

The piece is not all new, of course, just as the instrument upon which it is performed is composed of many parts, some as old (almost) as music itself—the strings and the bow, for example—and some utterly novel, like the sensors with which the bow is equipped. Machover has written that the musical seed from which the first ideas for sounds grew was the sarabande movement from Bach's second cello suite. With that ancestry, *Begin Again Again* ... forms a kind of historical essay, one that says: Look at what has happened to us since Master Bach showed us how it was to be done. Listen to how the realm of what can be considered as music has expanded; listen for what that expansion has done to our ability to perceive what is musical within that realm.

Faced with the unlimited freedom provided by his computer, that is, Machover has resorted to the construction of a kind of model—an idea in sound based on experience, the observation of beauty (meaning) contained within the Bach composition, through which to explore what happens to such beauty amid all those sounds now open to him. In Ma, the piece found a musician capable of asserting the presence of Bach within the accumulated layers of music and sound that have accreted since Bach's day. As performed, the piece established its own criteria, the standards by which it could be judged to have succeeded or failed to extract some meaning from the sea of data on which we float. Those two criteria are simply: Does it (the piece, the idea, the theory) work? Is it beautiful, does it reward us, does it please?

Not: Is it right, is it "true"? The standards for judging a piece of music, for judging whether a piece is "music" and not just noise, require subjective, relativistic measures: Does the work solve the problem at hand, can we hear it, does it communicate something to us; and do we like it? And as music is the art most like science, and as modern music makes its art out of the abstractions of information about sound, just as science elicits its results from data gleaned from nature—so the tests for whether or not any given idea in science makes sense turn on the same issues. To the extent that modern music does form an analogy to modern science, then science is not about truth as much as it is about its practical results, and about beauty.

The argument here is that the analogy between music and science is close enough to reveal the subjectivity, the limits on the claims of what science now tells us about experience. The problem is that in the midst of a rapidly changing era in science, the actual themes of our time that will seem as clear eventually as those of Pythagoras or Newton are obscured by the press of everyday work. Most scientists working short of the cutting edge of their fields (which is to say, most scientists), gather data and uncover phenomena that they know (correctly) to be "true." A protein sequence is that sequence, no ifs, ands, or buts; a pound of lead is both a pound and lead, unequivocally. Plenty of musicians could make the equivalent point for their work. Most of the music most of us hear virtually all the time are the uncomplicated tunes of popular compositions, built upon well-established, familiar chord patterns, the "truth" of old-fashioned harmony. John Lennon and Paul McCartney built many of their hits for the Beatles out of artfully, beautifully elaborated melodies and progressions through a relatively small number of different chords; they are simple—and they speak directly to their listeners.

To make matters worse, the informatization (to coin a horrible new word) of music guarantees the rapid propagation of much bad art, just as the use of models, and the corresponding manufacture of data, produces a wealth of incoherent, bad science. All the Muzak, all the tedious, endless "new age" knockoffs, all the grimly, relentlessly upbeat shimmer of advertisement sound tracks, can be produced that much faster, that much more thoughtlessly with the aid of the synthesizer. The technology of music has democratized the production of music, which means among other things that much more music can be made, which means, inevitably, that much of the music to be made will be absolutely banal. Similarly, the use of the computer and of models, those constructed representations of someone's idea of what nature looks like, ensures that the old GIGO law of computer science—Garbage In, Garbage Out—will remain in effect. And as models proliferate, what constitutes garbage expands beyond the original sense of it as bad, inaccurate data, faulty measurements fed into the computer program, to include bad ideas, poorly conceived models that can transform accurate inputs into meaningless results. The persistent failure of computer models to estimate the amount of the world's oil reserves, for example, represents a failure in the assumptions upon which such models are based rather than an error in the tabulation of current oil stocks.

But the certainty of much of science is a bit misleading, and the errors that accumulate within the information-based, modeled science are the unavoidable conseqence of a change in methods that has altered the fundamental nature of the scientific enterprise. Machover said that he truly understands the world as a mass of experience within which meaning lies; his synthesizers allow him to represent both the data and the structure to be found—or rather, a structure, one version, his—within that limitless expanse of possible experience. So it is with late-twentieth-century science. It is possible to know everything about a piece of the puzzle—your particular protein, the distribution of particles in one high-energy collision—but not all about the whole puzzle, not all at once. Instead, the universe as a whole manifests itself to our instruments as a sequence of data, never complete, never fully tallied. Machover was right: the world (not just music) can be seen as being made of information. As such it can be analyzed, processed. The computer or the abstract tool called a model—each plays the same universal role as the synthesizer does within music. And in all these cases, what those tools do for their users is the same: we gain an unbounded ability to ask the questions but must sacrifice our absolute confidence in our answers.

Which brings around again the issue of gauging the worth of what such inquiries turn up. Partly, as with music, the simplest test is whether or not the result works. Most of the time, this standard is sufficient: new results, new measurements, answer a question in such a way that the outcome leaves us with greater power over, or greater understanding of, our surroundings. Special relativity resolved a clearly evident paradox about the speed of light; a specific application of quantum mechanics led directly to the principle of the transistor. Both ideas work, and their success is their own, sufficient justification. But much, even most, of what people try to understand now is less amenable to such straightforward confirmation: we cannot predict the weather perfectly; we cannot account for the early life of the universe in complete detail; we do not understand why some smokers get cancer and others remain hale. So the test of an idea or an experimental result becomes a little weaker: it must be consistent, it must make sense. The one-mutation, one-tumor theory of cancer seemed logical but turns out to be false, and has been replaced with a model of cancer formation that involves a much more complex interaction between transformed cells and the immune system. That idea has yet to be fully confirmed, but it stands now, for it seems plausible at least, consistent with what is in fact understood.

But faced with such uncertainty, inherent in the construction of models of reality, there is one test more that can be given to a work of science, the same to which an audience puts any human attempt to bring order out of experience: Is the work beautiful? Einstein's most famous epigram was prompted by the question of what he would do if the measurements of bending starlight at the eclipse of 1919 contradict his general theory of relativity. He responded: "Then I would feel sorry for the good Lord. The theory is correct." It had to be, Einstein believed deeply, because it was too elegant, too beautiful to be false. More broadly, it has been an ongoing source of wonder that the universe is intelligible at all; that it seems to obey mathematical laws; that human reason can penetrate its structure even incompletely. But it seems, it has turned out, that the human ability to perceive structure, to recognize a quality we call beauty, to come to some agreement about what it might mean to be elegant, actually serves as a guide, a means of distinguishing (fallibly to be sure) between the gold and the dross in our ideas about the world.

To many ears the Machover piece worked by these two criteria, but not perfectly. It makes sense; its structure built up out of elements recognizably "really" music; it carries through; and it is dramatic and dynamic enough to give pleasure. At the same time, in it can be heard the effort and strain of the attempt to master a new medium. Similarly, the standard model of cosmology works, describing the evolution of the early universe in a manner consistent with observable data, but it, too, creaks a little, containing too many arbitrary numbers that have to be tuned to feel truly beautiful to those familiar with the mathematics. Hence Machover is at work on both new pieces and new instruments; many seek the cleaner, more efficient theory of everything of Steven Weinberg's dreams. The thread to draw out of the comparison of such incomparables is that both science and music, any art, are ultimately aesthetic endeavors. They give pleasure in the doing of them, and the proof of having done them well lies with the beauty and the coherence with which the outcomes of both present themselves to their audiences.

At this end of a long journey, such a thought must come as no surprise. It makes sense that music and science should resemble each other, for both tackle the same kind of task, resolving some image of human experience into an abstract form that can be communicated from one human being to another. The feature of science that throughout its history has marked it off from the other arts is the particular rigor imposed on it by the original discovery that the universe obeys mathematical laws. But the erosion of the authority that such rigor once granted to science caused by the discovery of uncertainty has in our own time restored that discipline to its place among the arts, the human crafts of order and abstraction. By the test of beauty we know that there is truth (or many, perhaps) about the world contained within Bach's cello suites still available to us after nearly three centuries. By the same tests we recognize the truths Pythagoras discovered, those Newton found, and so on, to the present day.

The essayist Kay Boyle pondered the question of what it takes to produce art, some distillation of experience that could convey not just facts but meaning from one mind to another. In one story she caught not only an example of the writer's art but also an explanation of what drives every attempt to perform such a feat in every discipline. In an essay on the teaching of writing she tells of a French resistance fighter, a friend of hers, who was caught and imprisoned in a Nazi concentration camp. Within the camp her friend and two or three other prisoners created a world. Every night, Boyle wrote, "they dressed for dinner in immaculate white shirts that did not exist, and placed, at times with some difficulty because of the starched material that wasn't there, pearl or ruby studs and cuff links in those shirts." Each evening the men would take on different identities and converse as befitted the people they became, over silverware, tablecloths, crystal, and candles that existed only in their focused imaginations. If the wine had not been properly corked on tasting, they sent it back. And "there were," Boyle reported, "certain restaurants they did not patronize a second time because the lobster had been overcooked."

Boyle's friend survived the camp, which is part of the point: he lived because in the phrase from Deuteronomy, he "chose life." But Boyle was writing about writing, not about the camps, and there is another meaning to be drawn from her parable. The resistance fighter found himself in a senseless world, a disordered, deadly one. By force of will, imagination, and artistry, he and his friends imposed on their inchoate experience an order, a sense of pattern, an argument about what their world could contain. It was not true, but as Boyle wrote: "The words they spoke were real." It worked; it made its own kind of sense; and each night's feast possessed an insane, glorious beauty.

FINALE:

DESCRIPTION

IS REVELATION

Science is something that human beings have always done to help them make sense of the world. Art, in its own way, aims at the same end. The revelation that emerges from this end of the twentieth century is that there is more to the story than that. Art and science don't just track the same quarry; they form between them a common endeavor, each presenting one face of the same impulse. Now, at the far end of 2,500 years of scientific thinking, exploring the link that binds the two shows why we still need both: the acts of making art and doing science don't just complement each other, they intertwine. We need each to help guide us through what will become an ever more complex future.

For the Greeks, of course, science and art cohabited as well: scientific truth made itself obvious by its beauty; artistry provided individual beautiful instances within which to recognize such truths. But if the idea seems the same as one held by our most distant antecedents, the context, of course, has altered irrevocably—and the same thought in a new setting takes on new meaning. We say, with the Greeks, that both art and science seek to describe, organize, and interpret the human encounter with our surroundings. But leaving the Greeks behind, we can now say: science behaves like other arts but is not identical to them. The most significant difference lies in the way modern science puts its results to the test. In spite of Greek (and romantic) claims, beauty isn't truth; rather, it is just one suggestive hint to the scientist that she is on the right track. The only, ultimate standard for the success of an idea in science is the sustained agreement between prediction and result. There are right and wrong answers (most of the time)—facts we can gather that describe something actually in the world.

And yet, remembering that science is a form of art enables us to see that it is not simply—or even so much—about nature out there as it is about ourselves, about making sense of that human condition. That's obvious enough in the so-called human sciences—everything from biology to psychology and maybe beyond. But all of science exists just as a work of music does, in the context of its audience, of those who look at its results and draw meaning from them; from those who listen, and respond. Science, even the most abstract of inquiries, remains absolutely a human passion. There is a recent story in science that illustrates the point. The astronomer George Smoot is one of the leaders of a long and very complex effort to measure cosmic background radiation—the residual heat of the big bang itself. That heat had been discovered (by accident) by two Bell Labs scientists in 1965. It was present wherever one looked from the earth, and everywhere it registered exactly the same: 5° F above absolute zero. That meant trouble—for if the big bang were perfectly uniform, it should by rights have produced a perfectly uniform universe—and yet there are quasars and galaxy clusters, galaxies, stars, and us to show that the universe in fact has a more complicated structure. The answer, cosmologists predicted, was that the background radiation did in fact vary, but only at very low levels, too difficult to measure from earth. So Smoot and his colleagues built an instrument to fly on a satellite, sent it up, got data back, analyzed it, and made a picture. The image they created was mottled and textured—a portrait of the big bang that showed that there were in fact tiny differences in energy levels within the universe at the earliest moments of its birth. There was a structure there, in the beginning, that would eventually lead to the existence of a star with a planet with people on it who could take its picture. Looking at that image, Smoot said, was like staring "at the face of God."

Well—not really. After all, that pattern of light and dark was what they had expected to see. The real surprise would have been if their supersensitive instrument, taking its data above the interference of the atmosphere, had still turned up a completely constant picture. That would have meant that the currently popular theories of cosmological origins had some deep problems: there would have been an unexplained gap between the events at the very beginning—the uniform big bang—and the evident, present reality of a nonuniform universe. Instead Smoot and his colleagues confirmed what the cosmologists predicted and hoped they would see, that the early universe possessed the structure needed to evolve into the mature one we inhabit. What Smoot had really glimpsed, perhaps, were the faces of some very relieved theorists.

But that's not quite right either, not a fair reading of Smoot's elation and the meaning of his discovery. The big bang itself happened (probably), giving birth to our mysteriously law-abiding universe. The *idea* of the big bang, though, is a creation, an artifact, something human beings have made to organize all those millennia of observations, the stray images of planets wandering in the sky, of stars and galaxies, of the strange fact that every place in the universe is rushing away from every place else and so on. Like any big idea in science—and like each successful work of art—cosmological theory makes a single, powerful statement about all that data, all the experience that led to the creating of the idea, to the making of the work of art. The joy Smoot found in looking at that motley snapshot of the big bang was the pleasure that comes when you realize that you have got it right. Most important—Smoot's response was aesthetic, not scientific. What he saw entranced him because in the context of what he knows, it was beautiful. Smoot and his colleagues possess a picture of how the world behaves, and with one, clear, simple, elegant result, they reaffirmed that their picture works. That's the motivation: the search for beauty, not in nature but in our representations of what we see, in our explanations of how the universe makes sense, how its facts organize themselves (how we organize the facts) into a coherent account. This is where science is closest kin to the arts.

We're back to the Greeks again, for whom the identification of an ordered phenomenon in nature served as evidence of the existence of Order generally, of the logic that governs the whole of experience. For us, each view of a universe that obeys the rules we have identified adds to our sense of understanding, of making sense of the whole. And yet once more, the distance we have traveled leaves us with a lesson different from the one the Greeks learned. Since the scientific revolution, science has been seen as a tool of mastery, one that will enable us to predict, even manipulate, nature, while we stand, in a sense, outside of nature. We know now that isn't so; we cannot know all we need to know, all we want to know. But if we recall that science is an art, then we can take the measure of the strength that lies within this apparently diminished role for scientific discovery. A science that can provide a glimpse of the face of (some kind of) God is one that can help us come to terms with our existence within the world—to understand and not simply to command our circumstances.

What does science do? What can scientists provide; and what should we seek from them, at this far end of a journey we began by asking Why? The short, bald answer is that science quantifies our perceptions of reality; in doing so it forms a way of seeing that enables us to measure not just nature but ourselves. Wallace Stevens, who constantly prowled this borderland between art and science, put it more sharply in a few lines from "Description Without Place":

> *Description is revelation. It is not*
> *The thing described, nor false facsimile.*
>
> *It is an artificial thing that exists,*
> *In its own seeming, plainly visible,*
>
> *Yet not too closely the double of our lives,*
> *Intenser than any actual life could be,*
>
> *A text we should be born that we might read,*
> *More explicit than the experience of the sun.*

We look; we consider what we see; we create—not find, not discover, but create—a portrait of what we have seen. It is not *the* truth, but *a* truth, one that is neither the thing described nor its false facsimile. It is real—we have made it. It distills our experience, "intenser than any actual life." We grasp for the knowledge that enables us to describe, for in description we seek and sometimes find revelation. The ultimate revelation is that which leads us to ourselves, to some sense of who we are, where (in the largest sense) we live, how we have come to this pass. In times past the learned referred to "the book of nature." The image is apt, for as we study nature we write that book and then strive to understand what our discoveries have written. That text is still growing. It is one that we are born to read.

ACKNOWLEDGMENTS

This book has gained enormously from the help and support of a shockingly large number of busy people. Of them all, two have been at the forefront. My agent, Sallie Gouverneur, has once again helped to shape this project from beginning to end. My editor, Rebecca Saletan, has performed as few in publishing still do, helping me wrestle with my subject, and then wrestling in her turn to make what I said as clear and as elegant as possible. Without them this book would never have happened, and to them my deepest thanks.

I am indebted to everyone whose work appears in the pages that follow. Some gave so much that I must thank them again, publicly. Yo Yo Ma worked hard to convey the experience of making music on a Stradivarius cello. Ma's manager, Mary Pat Buerkle of ICM, and Ma's assistants, Bonnie Dane Evans and Sarah Stackhouse, worked hard to open the windows of time in Ma's insane schedule for me. Wendy and Peter Moes showed me some part of what it takes to build first-class instruments. My interchange with Michael McCune on his chimerical mouse work has gone on for several years. Tod Machover and Michael Hawley, both of MIT's Media Lab, each gave me repeated interviews to convey some sense of their work at the cutting edge of musical technology. The Marine Biological Laboratory's Shinya Inoue introduced me to the history and his own remarkable current practice of optical microscopy over a scattered decade of conversations.

Many more generously offered time, effort, and patience to help guide me through the complexities of both music and science. Greg Bover of the C.B. Fisk Company laid bare the process of organ building for me. Laurie Monaghan together with her colleagues in the Ensemble Project Ars Nova introduced me to the world of medieval music performance. Winold Van Der Putten's work reconstructing Henri Arnaut's portative organs was invaluable to me, as were conversations with him and his colleague in organ archaeology, Jankaes Braaksema. Charles Beare of J. and A. Beare, Ltd., shared his expert knowledge of both Stradivarius and the history of the violin family. Stephen Schneider of Stanford University has helped me think about models in science for almost a decade. Martijn Sanders, director of the Concertgebouw in Amsterdam, and his assistant, Marijke Van Oordt, showed me extraordinary hospitality throughout a remarkable week of concerts at what is probably the best—and best run—concert hall in the world.

Several people who should have known better agreed to read my work. They have made it significantly better than it was on its own; the faults and errors that remain are all mine. Professor Nathan Sivin of the University of Pennsylvania's History and Sociology of Science department reviewed the entire book, and his comments were invaluable. Philharmonia Baroque conductor Nicholas McGegan provided his musician's perspective on the entire text, which was again enormously helpful. MIT music history professor Lowell Lindgren saved me from the worst of my sins of omission and commission in the four music chapters, as did David Tayler of the University of California at Berkeley. Nobel Laureate chemist Roald Hoffman read the sections on alchemy and atomism, conveying his great love and admiration for the world of magic that helped shape his world of science. Dr. Joy Hirsch, director of the Memorial Sloan Kettering Hospital vision research laboratory, reviewed the portions concerned with optics, telescopes, and microscopes. Finally, Boston's Professor Theoharis Constantine Theoharis followed the book in progress and meticulously read the completed manuscript, serving as both mentor and midwife throughout the project—to him a special thanks.

My gratitude as well to the members of Simon & Schuster's staff who have done yeoman work to make this book happen. Assistant editor Denise Roy kept both her boss Saletan and me on track. Production editor Jay Schweitzer kept what became a complicated project moving smoothly through to publication. Bruce Macomber gave the book a sensitive and expert copyedit; if the finished product is lucid, much of the credit goes to him.

Acknowledgments

Finally, Frank and Eve Metz, in charge of the exterior and interior design of the book respectively, took an interest in the project early on, and have given me enormous pleasure by their efforts to make this book beautiful, as have cover designer Karin Goldberg and book designer Pei Loi Koay. Jane Callander, London based picture researcher, did expert work swiftly and resourcefully. Johanna Bartelt aided me in the trying process of gathering picture permissions.

Lastly, friends and family have made the long process of writing both tolerable and—when I was either lucky or good—a joy. Some who listened far longer than they had to include—in no order—Steven Latham, Alex Shapiro, Jon Eckstein, Kelly Roney, David Vanderburgh, Brett Oberman, Eleanor Powers, Sean Vickery, Laura Besvinick, Paul Andreasson, Gary Taubes, Dennis Overbye, Peter Jones, Paula Apsell, Bill Grant, Evan Hadingham, Joseph Levine, Gino del Guercio, Stephanie Munroe, Boyd Estus, Noel Schwerin, Noel Buckner, Rob Whittlesey, and Merry White. My ultimate and heartfelt thanks to my extended family—Dan and Helen Levenson, their daughters Rachel, Marilyn and Judy; Jack, Charlotte, Paul and Laurie Levenson; my siblings Richard, Irene, and Leo, their spouses Jan, George, and Kathy; and to my mother Rosemary Levenson.

ILLUSTRATION

CREDITS

Illustrations appear courtesy of the following sources:

Pages 20 and 187: Scala/Art Resource, NY

Page 23: Bayer. Staatsbibliothek, Munich

Pages 25, 45, 47, 49, 89, 91, 94, 97, 122, 125, 137, 244, and 279; By permission of the Houghton Library, Harvard University

Page 31: University Library, University of Utrecht

Page 60: Museo Civico Medievale

Page 66: Musée Condé

Page 84: The Metropolitan Museum of Art, H. O. Havemeyer Collection, bequest of Mrs. H. O. Havemeyer, 1929. (29.100.5)

Pages 102 and 105: By permission of the President and Council of the Royal Society, London

Pages 144 and 148: The Warburg Institute

Pages 179 and 281: The Science Museum/Science and Society Picture Library, London

Pages 209 and 211: Libraria del Convegno

Page 214: The Catgut Acoustical Society

Page 251: The Cleveland Library

Page 294: NYT Pictures

Page 297: Stockhausen-Verlag

Page 305: Philips Consumer Electronics BD

BIBLIOGRAPHY

CHAPTER 1

S. Sambursky's book is the classic account of Greek knowledge and beliefs about science. Flora Levin's monograph is a good introduction to Pythagorean music theory, and Jean Perrot's book on the organ is a fine treatment of the history of that instrument.

Atwater, Donald. *The Penguin Dictionary of the Saints*. 2d ed. London: Penguin, c. 1983.

Audsley, George Ashdown. *The Art of Organ Building*. 2 vols. New York: Dover, 1965.

Augustine of Hippo. Translation and Introduction by Mary T. Clark. Mahwah, N.J.: Paulist Press, 1984.

Bernal, Jacques. *Science in History*. 4 vols. Cambridge: MIT Press, 1971.

Bernstein, Leonard. *The Unanswered Question*. Cambridge: Harvard University Press, 1976.

Burkert, Walter. *Lore and Science in Ancient Pythagoreanism*. Cambridge: Harvard University Press, 1972.

Ferrero, Guglielmo. *The Life of Caesar*. New York: Norton, 1962.

Gibbon, Edward. *The Decline and Fall of the Roman Empire*. Everyman's Library, 1910.

Lavalleye, Jacques. *Memling à l'hôpital St. Jean*. Brussels: Charles Dessart, 1953.

Levin, Flora. *The Harmonics of Nichomachus and the Pythagorean Tradition*. American Classical Studies no. 1. University Park, Pa.: American Philological Association, 1975.

Perrot, Jean. *The Organ*. London: Oxford University Press, 1971.

Plato. *Timaeus and Critias*. London: Penguin, 1977.

The Pythagorean Sourcebook and Library. Compiled and translated by Kenneth Sylvan Guthrie; edited by David R. Fideler. Grand Rapids, Michigan: Phanes Press, 1987.

Sambursky, S. *The Physical World of the Greeks*. Princeton: Princeton University Press, 1956.

Usher, Abbot Payson. *A History of Mechanical Inventions.* Rev. ed. New York: Dover, 1982.

The Work of Hans Memling. New York: Brentano's, 1913 (no author).

CHAPTER 2

John Fauvel, et al.'s *Let Newton Be* is an excellent introduction to the wide range of Newton's work. In addition to the essay on Newton's musical interests, this collection includes essays on Newton's optics, his philosophy, and his interest in alchemy, all with good lists of suggested additional reading. Morris Kline's works are excellent, and not enough read, and Hoppin gives a clear and concise introduction to the incredible complexity of medieval music. For the words of medieval theorists themselves, please consult Strunk's very useful compilation. The Schiller Institute's *Manual* is included as a kind of curiosity—with an introduction by Lyndon Larouche, it attempts to link the decline of the West to the abandonment of well-tempered tuning systems.

Abraham, Gerald. *The Concise Oxford History of Music.* Oxford: Oxford University Press, 1979.

Apel, Willi. *Gregorian Chant.* Bloomington: Indiana University Press, 1958.

———. *Medieval Music.* Stuttgart: F. Steiner Verlag, Wiesbaden GMBH, 1986.

———. *The Notation of Polyphonic Music.* Cambridge: Medieval Academy of America, 1953.

Barber, Richard. *The Penguin Guide to Medieval Europe.* London: Penguin, 1984.

Castiglione, Baldesar. *The Book of the Courtier.* Translated by George Bull. London: Penguin, 1967.

Eliot, T. S. *Four Quartets.* New York: Harcourt, Brace, 1943.

Fauvel, John; Raymond Flood; Michael Shortland; and Robin Wilson, eds. *Let Newton Be.* Oxford: Oxford University Press, 1988.

Geiringer, Karl. *Johann Sebastian Bach.* Oxford: Oxford University Press, 1966.

Helmholtz, Hermann. *On the Sensations of Tone.* New York: Dover, 1954.

Herz, Gerhard. *Essays on J.S. Bach.* Ann Arbor: UMI Research Press, 1985.

Hoppin, Richard H. *Medieval Music.* New York: Norton, 1978.

Jeans, James. *Science and Music.* New York: Dover, 1968.

Kline, Morris. *Mathematics in Western Culture.* Oxford: Oxford University Press, 1953.

———. *Mathematics and the Search for Knowledge.* Oxford: Oxford University Press, 1985.

Le Goff, Jacques. *Medieval Civilization.* Oxford: Basil Blackwell, 1988.

A Manual on the Rudiments of Tuning and Registration. Washington, D.C.: Schiller Institute, 1992.

Strunk, W. Oliver. *Source Readings in Music History from Classical Antiquity through the Romantic Era.* New York: Norton, 1950.

Terry, Charles Sanford. *Bach.* 2d ed. London: Oxford University Press, 1933.

Williams, Peter, and Barbara Owen. *The Organ.* New York: Norton, 1980, 1984 (The New Grove Musical Instruments Series).

Williams, Peter. *A New History of the Organ.* Bloomington: Indiana University Press, 1980.

Yudkin, Jeremy. *Music in Medieval Europe.* New York: Prentice Hall, 1989.

CHAPTER 3

Galileo and Hooke's books are two of the most readable original works in the literature of science. Ford's study, *Single Lens,* is a comprehensive introduction to Leeuwenhoek's work. Lindberg's treatment of Bacon in his *The Beginnings of Western Science* provides an excellent account of Bacon in the context of his own time and ambitions. There are no single sources that would provide a comprehensive portrait of the non-Western scientific enterprise and its interplay with developments in Europe. Good places to begin filling in that largely neglected and important story include Joseph Needham's encyclopedic *Science and Civilization in China,* and for Islamic contributions, please see Thomas Goldstein's work, and that of A. I. Sabra, represented below as one of the contributors to Bernard Lewis's *The World of Islam.*

Bacon, Roger. *The Mirror of Alchimy.* Edited by Stanton J. Linden. New York: Garland Publishers, 1992.

———. *On the Nullity of Magic.* New York: AMS Press, 1923.

———. *The Opus Majus of Roger Bacon.* 3 vols. Oxford: Clarendon Press, 1897–1900.

Biagioli, Mario. *Galileo, Courtier.* Chicago: University of Chicago Press, 1993.

Clay, Reginald S., and Thomas H. Court. *The History of the Microscope.* New York: Longwood Press, 1978.

Crombie, A. C. *Augustine to Galileo.* 2 vols. London: Mercury Books, 1961.

———. *Science, Optics and Music in Medieval and Early Modern Thought.* London: Hambledon Press, 1990.

Draxe, Stillman. *Galileo.* Oxford: Oxford University Press, 1980.

Finocchiaro, Maurice A. *The Galileo Affair.* Berkeley: University of California Press, 1989.

Ford, Brian J. *Single Lens.* New York: Harper & Row, 1985.

Galilei, Galileo. *Sidereus Nuncius.* Translated by Albert Van Helden. Chicago: University of Chicago Press, 1989.

Goldstein, Thomas. *Dawn of Modern Science.* Boston: Houghton Mifflin, 1980.

Hooke, Robert. *Micrographia.* Early Science in Oxford series, vol. 13. Edited by R. T. Gunther. Oxford: Oxford University Press, 1938.

King, Henry C. *The History of the Telescope.* New York: Dover, 1955.

Lewis, Bernard. *The World of Islam.* London: Thames and Hudson, 1976.

Lindberg, David. *The Beginnings of Western Science.* Chicago: University of Chicago Press, 1992.

———. *Roger Bacon's Philosophy of Nature.* New York: Oxford University Press, 1983.

Manuel, Frank E. *A Portrait of Isaac Newton.* Cambridge: Harvard University Press, 1968.
Needham, Joseph. *Science and Civilization in China.* Cambridge: Cambridge University Press, 1954–1988.
Newton, Isaac. *Opticks.* New York: Dover, 1952.
Ronchi, Vasco. *Optics.* New York: Dover, 1991.
Sabra, A. I. *Theories of Light from Descartes to Newton.* Cambridge: Cambridge University Press, 1981.
Singer, Charles. *From Magic to Science.* London: Ernest Benn, 1928.
Turner, G. L'E. *Essays on the History of the Microscope.* Oxford: Senecio Publishing, 1980.
Welford, W. T. *Optics.* 3d ed. Oxford: Oxford University Press, 1988.

PART 1 INTERLUDE

Both of the classic biographies of Bach contain a great deal of information; Spitta's is perhaps slightly more comprehensive. Hawking's bestseller contains a good reminder that theological ambition remains alive and well for at least some of that most ambitious of intellectual tribes, the theoretical physicists. Bronowski's book is a delight.

Bronowski, J. *Science and Human Values.* New York: Harper & Row, 1975.
Hallyn, Fernand. *The Poetic Structure of the World.* New York: Zone Books, 1990.
Hawking, Stephen. *A Brief History of Time.* New York: Bantam Books, 1988.
Heisenberg, Werner. *Physics and Beyond.* New York: Harper & Row, 1972.
Kepler, Johannes. *The Harmonies of the World.* Great Books of the Western World, vol. 16. Chicago: Encyclopedia Britannica, 1955.
Schweitzer, Albert. *J. S. Bach.* 2 vols. Boston: Bruce Humphries, 1911.
Spitta, Phillip. *Johann Sebastian Bach.* 3 vols. New York: Dover, 1951.

CHAPTER 4

Brian Vickers's collection is an excellent review of modern scholarly thinking about the history of the occult element in natural science—see especially his essay. Ernst Cassirer's *The Renaissance Philosophy of Man* contains a translation of Pico della Mirandola's Oration on the Dignity of Man. Lynn Thorndike's massive work contains reference to virtually every western thinker who speculated on magic, science, and the two together. It is hard to use, but comprehensive.

Cassirer, Ernst. *The Philosophy of the Enlightenment.* Boston: Beacon Press, 1955.
Cassirer, Ernst, et al., eds. *The Renaissance Philosophy of Man.* Chicago: University of Chicago Press, 1948.
Dobbs, Betty Jo Teeter. *The Janus Faces of Genius.* Cambridge: Cambridge University Press, 1991.
Greenburg, Sidney Thomas. *The Infinite in Giordano Bruno.* New York: Octagon Books, 1978.

Hanson, P. M., and Alan E. Shapiro. *The Investigation of Difficult Things.* Cambridge: Cambridge University Press, 1992.

Harré, Rom, ed. *The Physical Sciences Since Antiquity.* London: Croom Helm, 1986.

Harrison, John. *The Library of Isaac Newton.* Cambridge: Cambridge University Press, 1978.

Holmyard, Eric John. *The Makers of Chemistry.* Oxford: Oxford University Press, 1931.

Hopkins, Arthur John. *Alchemy, Child of Greek Philosophy.* New York: Columbia University Press, 1934.

Jonson, Ben. *The Alchemist.* Manchester: Manchester University Press, 1979 (originally William Stansby, printer, London, 1616).

Lindsay, Jack. *The Origins of Alchemy in Graeco-Roman Egypt.* London: Frederick Muller, 1970.

Luck, Georg. *Arcana Mundi.* Baltimore: Johns Hopkins University Press, 1985.

Sadoul, Jacques. *Alchemists and Gold.* New York: G. P. Putnam and Sons, 1972.

Thorndike, Lynn. *A History of Magic and Experimental Science.* 8 vols. New York: Columbia University Press, 1941.

Vickers, Brian, ed. *Occult and Scientific Mentalities in the Renaissance.* Cambridge: Cambridge University Press, 1984.

Walker, D. P. *Unclean Spirits.* London: Scolar Press, 1981.

Webster, Charles. *From Paracelsus to Newton.* Cambridge: Cambridge University Press, 1982.

Yates, Frances A. *Giordano Bruno and the Hermetic Tradition.* Chicago: Chicago University Press, 1964.

CHAPTER 5

The *Annals of Philosophy* is the English language source for Berzelius and much other early nineteenth-century chemical writing. Harré's book is an excellent introduction to the methods and aims of experimental science and includes descriptions of both Lavoisier's and Berzelius's key works. Knight's essay is a good treatment of the scientific interest in organizing sense experience.

Berzelius, Jöns Jakob. *Autobiographical Notes.* Translated by Olof Larsell. Baltimore: Williams & Wilkins, 1934.

Cohen, I. B. *Revolution in Science.* Cambridge: Harvard University Press, 1985.

Farber, Eduard. *The Evolution of Chemistry.* 2d ed. New York: Ronald Press, 1969.

Fox, Robert. *The Caloric Theory of Gases.* Oxford: Clarendon Press, 1971.

Guerlac, Henry. *Lavoisier—The Crucial Year.* New York: Gordon and Breach, 1961.

Hall, A. Rupert. *Isaac Newton.* Oxford: Blackwell Publishers, 1993.

———. *The Scientific Revolution.* Boston: Beacon Press, 1954.

Hartley, Harold. *Studies in the History of Chemistry.* Oxford: Clarendon Press, 1971.
Harré, Rom. *Great Scientific Experiments.* Oxford: Oxford University Press, 1983.
Ihde, Aaron J. *The Development of Modern Chemistry.* New York: Dover, 1964.
Jorpes, J. Erik. *Jacob Berzelius.* Stockholm: Almqvist & Wiksell, 1966.
Kisch, Bruno. *Scales and Weights.* New Haven: Yale University Press, 1965.
Klein, Herbert Arthur. *The Science of Measurement.* New York: Dover, 1974.
Knight, David. *Ordering the World.* London: Burnett Books, 1981.
Lavoisier, Antoine. *Traité Elémentaire de Chimie.* Translated by Robert Kerr. Edinburgh: William Creech, 1790; reprinted by Edwards Brothers, Ann Arbor, 1945.
Melhado, Evan, and Tore Frängsmyr, eds. *Enlightenment Science in the Romantic Era.* Cambridge: Cambridge University Press, 1992.
Melhado, Evan. *Jacob Berzelius.* Madison: University of Wisconsin Press, 1981.
Moody, Ernest A., and Marshall Clagett. *The Medieval Science of Weights.* Madison: University of Wisconsin Press, 1952.
O'Brien, D. *Weight in the Ancient World.* 2 vols. Paris: Les Belles Lettres, 1981.
Partington, J. R. *A History of Chemistry.* New York: Macmillan, 1972.
Rocke, Alan J. *Chemical Atomism in the Nineteenth Century.* Columbus: Ohio State University Press, 1984.
Shapin, Steven. *Leviathan and the Air Pump.* Princeton: Princeton University Press, 1985.
Sydenham, P. H. *Measuring Instruments.* Stevenage, U.K.: Peter Peregrinus, 1979.
Thackray, Arnold. *Atoms and Powers.* Cambridge: Harvard University Press, 1970.
Thomson, Thomas, ed. *The Annals of Philosophy.* Vols. 2–15; new series, vol. 9. Robert Baldwin, 1813–1825.
Warren, Charles. *The Early Weights and Measures of Mankind.* London: Committee of the Palestine Exploration Fund, 1913.

CHAPTER 6

There are many books on nonlinear dynamics and chaos. James Glieck's bestseller is a good popular introduction, and Ivar Ekeland's is a delight to read. David Bohm's book is different. It contains an idiosyncratic and controversial account of the claims made for physical law, and it provides a clear and eloquent discussion of the gap between physical law and the material circumstances of the real world. Maynard Solomon's book contains a translation of Beethoven's *Tagebuch,* his journal.

Bohm, David. *Causàlity and Chance in Modern Physics.* Philadelphia: University of Pennsylvania Press, 1987.
Boyden, David D., et al. *The Violin Family.* New York: Norton, 1980.

Campbell, Margaret. *The Great Cellists*. North Pomfret, Vt.: Trafalgar Square Publishers, 1989.

Cohen, H. F. *Quantifying Music*. Dordrecht: D. Reidel, 1984.

Cowling, Elizabeth. *The Cello*. New York: Charles Scribner's Sons, 1975.

Doring, Ernest N. *How Many Strads?*. Chicago: William Lewis & Son, 1945.

Ekeland, Ivar. *Mathematics and the Unexpected*. Chicago: University of Chicago Press, 1988.

Fletcher, N. H., and T. D. Rossing. *The Physics of Musical Instruments*. New York: Springer Verlag, 1991.

Hall, Donald E. *Musical Acoustics*. 2d ed. Pacific Grove: Brooks/Cole Publishing Company, 1991.

Hart, G. *The Violin*. London: Dulao, 1885.

Hill, W. Henry, Arthur F., and Alfred E. *Antonio Stradivari*. London: Macmillan, 1909.

Markevitch, Dimitry. *The Cello Story*. Princeton: Summy-Birchard Music, 1984.

Mellers, Wilfred. *Bach and the Dance of God*. New York: Oxford University Press, 1981.

Pleeth, William. *Cello*. New York: Schirmer Books, 1982.

Sacconi, Simone F. *The "Secrets" of Stradivari*. Cremona: Libreria del Convegno, 1979.

Solomon, Maynard. *Beethoven Essays*. Cambridge: Harvard University Press, 1988.

The Tuscan and Le Messie. London: W. E. Hill and Sons, 1891, 1976.

Vogt, Hans. *Johann Sebastian Bach's Chamber Music*. Portland: Amadeus Press, 1988.

Wolff, Christoph, et al. *The New Grove Bach Family*. New York: Norton, 1983.

PART 2 INTERLUDE

Poincaré is an underappreciated commentator on science, as well as having been one of its truly great practitioners. Each of his collections of essays contains insights that still provoke and inform. Stravinsky speaks as well for himself as any of his interpreters since have spoken for him. If he is a biased commentator, that bias comes along with a passionate commitment to a conception of music laid out quite clearly in his Norton Lectures volume, *The Poetics of Music*.

Albright, Daniel. *Stravinsky*. New York: Gordon and Breach, 1989.

Andriessen, Louis, and Elmer Schönberger. *The Apollonian Clockwork: On Stravinsky*. Translated by Jeff Hamburg. Oxford: Oxford University Press, 1989.

King-Hele, D. G., and A. R. Hall. *Newton's Principia and Its Legacy*. London: The Royal Society, 1988.

Laplace, Pierre. *The System of the World*. 2 vols. Translated by Henry Harte. London: Longman, Rees, Orme Brown, and Green, 1830.

Pasler, Jann, ed. *Confronting Stravinsky*. Berkeley: University of California Press, 1986.

Poincaré, Henri. *Science and Hypothesis.* New York: Dover, 1952.
———. *The Value of Science.* New York: Science Press, 1907.
———. *Science and Method.* New York: Dover, 1952.
Stravinsky, Igor. *The Poetics of Music.* Cambridge: Harvard University Press, 1970.
Whitman, Walt. *The Poetry and Prose of Walt Whitman.* Edited by Louis Untermeyer. New York: Simon & Schuster, 1949.

CHAPTER 7

SCID-hu mouse research is too new to have made it to a book form—all its documents are papers, published in the technical literature. "The SCID-hu Mouse: A Small Animal Model for HIV Infection and Pathogenesis" in the *Annual Review of Immunology* 9 (1991): 399–429, McCune et al., is a good overview of work on the system. Richardson's book offers a first statement of the aims of model building, and Schneider and Londer in their sprawling work include one succinct, clear section that details how climate and weather models work. Lorenz, the first to demonstrate the inherent unpredictability of the weather, provides an account of the problem faced by model builders confronting the atmosphere—and by extension, those confronting model builders seeking to understand any complex, nonlinear system.

Aspray, William. *John Von Neumann and the Origins of Modern Computing.* Cambridge: MIT Press, 1990.
Calvino, Italo. *Invisible Cities.* New York: Harcourt Brace Jovanovich, 1974.
Keegan, John. *Six Armies in Normandy.* New York: Viking, 1982.
Levy, Steven. *Artificial Life.* New York: Pantheon, 1992.
Lorenz, E. *The Nature and Theory of the General Circulation of the Atmosphere.* Geneva: World Meteorological Association, 1967.
Macrae, Norman. *John Von Neumann.* New York: Pantheon, 1992.
Plato. *The Republic.* New York: Charles Scribner's Sons, 1928.
Rhodes, Richard. *The Making of the Atomic Bomb.* New York: Simon & Schuster, 1986.
Richardson, L. F. *Weather Prediction by Numerical Process.* Cambridge: Cambridge University Press, 1922.
Root-Bernstein, Roger. *Rethinking AIDS.* New York: Free Press, 1993.
Schneider, Stephen, and Randi Londer. *The Coevolution of Climate and Life.* San Francisco: Sierra Club Books, 1984.
Shore, William, ed. *Mysteries of Life and the Universe.* New York: Harcourt Brace Jovanovich, 1992.
Weinberg, Steven. *Dreams of a Final Theory.* New York: Pantheon, 1992.

CHAPTER 8

Jon Appleton's short essay offers a good glimpse of what a professional composer seeks from digital instruments. Robert Lucky's *Silicon*

Dreams is the best lay introduction to information science I have encountered. *The Encyclopedia of Automatic Musical Instruments* is a soup-to-nuts guide to the field. There are several good books on the techniques and aims of computer music. Charles Dodge's and Thomas Jerse's work was particularly useful. Edward Cone's brief study of musical performance remains one of the clearest attempts to analyze how a performer and the work interact to create art.

Appleton, Jon. *21st Century Musical Instruments*. New York: Institute for Studies in American Music, 1989.

Bartók, Béla. *Essays*. London: Faber and Faber, 1976.

Boyle, Kay. *Words That Must Somehow Be Said*. San Francisco: North Point Press, 1985.

Campbell, Jeremy. *Grammatical Man*. New York: Simon & Schuster, 1982.

Cone, Edward T. *Musical Form and Musical Performance*. New York: Norton, 1968.

Darter, Tom, compiler; Greg Armbruster, ed. *The Art of Electronic Music*. New York: Quill, 1984.

Dodge, Charles, and Thomas A. Jerse. *Computer Music*. New York: Schirmer Books, 1985.

The Encyclopedia of Automatic Musical Instruments. Vestal, N.Y.: Vestal Press, 1972.

Ernst, David. *The Evolution of Electronic Music*. New York: Schirmer Books, 1977.

Holmes, Thomas B. *Electronic and Experimental Music*. New York: Charles Scribner's Sons, 1985.

Horn, Delton T. *Digital Electronic Music Synthesizers*. 2d ed. Blue Ridge Summit, Pa.: Tab Books, 1988.

Lucky, Robert. *Silicon Dreams*. New York: St. Martin's Press, 1989.

Maconie, Robin. *The Works of Karlheinz Stockhausen*. Oxford: Clarendon Press, 1990.

Mann, Thomas. *Doctor Faustus*. New York: Knopf, 1948.

Manning, Peter. *Electronic and Computer Music*. Oxford: Clarendon Press, 1985.

Moore, F. Richard. *Elements of Computer Music*. Englewood Cliffs, N.J.: Prentice Hall, 1990.

Ord-Hume, Arthur W. J. G. *Pianola*. London: George Allen & Unwin, 1984.

Roads, Curtis, ed. *Composers and the Computer*. Los Altos, Calif.: William Kaufmann, 1985.

———. *The Music Machine*. Cambridge: MIT Press, 1989.

Roehl, Harvey. *The Player Piano*. Watkins Glen, N.Y.: Century House, 1958.

Rothstein, Joseph. *MIDI: A Comprehensive Introduction*. Madison: A-R Editions, 1992.

Rumsey, Francis. *MIDI Systems and Control*. London: Focal Press, 1990.

Schoenberg, Arnold. *Style and Idea.* Berkeley: University of California Press, 1984.

Schaeffer, Pierre. *A la Recherche d'une musique concrète.* Paris: Editions de Seuil, 1952.

Stockhausen on Music. Compiled by Robin Maconie. London: Marion Boyars, 1989.

Tannenbaum, Mya. *Conversations with Stockhausen.* Oxford: Clarendon Press, 1987.

INDEX

Abbey, Joseph, 59
abstraction, 232, 312
 in mathematics, 247
 in models, 229, 261, 269
 in music, 237–40
acoustical physics, 211–15, 220,
 224, 226
Aelred, Saint, 36, 38, 53
aesthetics, 114–16, 309, 311–13,
 315–18
*Afhandlingar i Fysik, Kemi, och Min-
 eralogi* (*Transactions of Physics,
 Chemistry, and Mineralogy*),
 156
Agricola de Metallica (*The Farming of
 Metal*), 122
AIDS research:
 direct, 269
 SCID-hu mouse in, 243–45, 249,
 260–70
Aineas of Gaza, 125–26
air, *see* atmosphere
Alaric I, king of Visigoths, 33
Alchemist, The (Jonson), 143–44, 149
alchemy, 14, 121–35, 142–54, 161,
 172
 in ancient Egypt, 125–32
 bottom-up approach in, 124, 126,
 131, 134
 calces produced in, 168
 chemical atom and, 157
 commercial spirit and, 143–44
 distilled liquor invented in,
 134–35
 experiments conducted in, 123,
 124, 133–34
 as forerunner of chemistry, 143,
 145–46, 147–48, 153, 157, 166

Hermes as patron saint of, 131–32
Islamic pursuit of, 132–35
laboratory apparatus of, 128–31
literature of, 122, 130–31
medicine and, 134, 145–49, 165
microcosm-macrocosm unity in,
 131–32, 134, 146–47, 148–49,
 150–52, 249
in Middle Ages, 132, 133,
 134–35
natural magic and, 135–36, 143
natural processes replicated in,
 130–31, 132
natural sciences and, 150–51
nature as viewed in, 124, 126,
 130–31, 132, 134, 135–36, 151,
 152
Newton's study of, 121–24, 125,
 150, 155–56
Newton's worldview and,
 150–54
officialdom's suppression of,
 126–27
power derived from, 132, 152–53
precise weight measurement
 and, 165
radioactivity and, 189
recipes in, 127
in Renaissance, 121–24, 127,
 132, 135, 142–54
subsumed by science, 156
Alcides, 30
alcohol, distillation of, 134–35
algebra, 33
alloys, 126
"Amans ames secretement"
 ("Lovers, Love Secretly")
 (Cordier), 65–66

organ (*cont.*)
in Winchester, England, 50–51,
52, 54
Organ, The (Perrot), 30
organum, 48–49, 54–55
Oscar II, king of Sweden, 234
Othello (Shakespeare), 13
overtones, 26–27
of cello, 206–7
oxygen, 169, 170, 172
in air, 175
isotopes of, 189
metallic oxides and, 181
in water, 177

Padua, University of, 85, 86
Paganini, Niccolò, 203
Paolo da Firenze, 63
Paracelsus, Philippus Aureolus
(Theophrastus Bombastus von
Hohenheim), 135, 144–49, 167
attacks on, 148–49
disease model of, 146–47
medicine transformed by, 144–
146, 147–48
Paris, University of, 72
Aristotelian learning banned at,
75–76
Bacon's lectures at, 72, 76–77
Parker, Daniel, 215–16
Paul, Les, 295
pentatonic scale, 27
Pepin III, king of Franks, 34, 42
Pepys, Samuel, 98
perfection, 20, 21, 37
ascribed to heavenly bodies, 88,
90, 111
experienced through organ, 32
implicit in number "three," 64
musical, blows to concept of,
55–61, 62
in musical performance, 231, 232
music as expression of, 43–44,
64, 68
square notation and, 64
Perrot, Jean, 30, 31, 34
Persia, 33
Peter, Saint, 41
Petrarch, 82
Philip, duke of Burgundy, 57, 68
Philo, 30
Philolaus, 32, 36

philosopher's stone, 123, 132, 144,
149, 150, 154
phlogiston, 166–67, 168–69, 170,
171
physics:
acoustical, 211–15, 220, 224,
226
final theory sought in, 247–49,
272, 312
musical issues and, 197
quantum mechanics and, 190,
228, 258, 311
worldview of scientific revolution
ruptured by, 189–90
piano:
recording performances on, 286–
288, 292
sound production in, 26–27
see also player piano
Pianola (Ord-Hume), 284
Piatti, Alfredo, 205
Pico della Mirandola, Giovanni,
138–40, 152
pitch:
frequency and, 21
harmonious conjunction of number and, 21–24
notation of, 45, 47
plague, 134
planets:
Ficino's talismans and, 136–38
Galileo's observations of, 88–93,
109–10
music of, 13, 24, 112, 246
motion of, 24, 93, 94, 106, 110–
112, 114, 225, 234, 246
Plato, 271–72
player piano, 283–91, 292–93
audience-music relationship
transformed by, 287–88
Bartók's views on, 289–90
capabilities of mere mortals ex-
ceeded with, 288, 289
composers liberated from performers' limitations by, 288–
289, 290–91
electronic instruments compared
to, 298
Hawley's computer-controlled
Bösendorfer, 274–75, 288–89
nature of performance altered
by, 289–91, 298